D0461603

—THE NEW—
RIDER'S
COMPANION

Edited by Emma Callery

THE NEW
RIDER'S
COMPANION

Edited by Emma Callery
Photographs by Bob Langrish

CHARTWELL
BOOKS, INC.

Published by Chartwell Books
A Division of Book Sales Inc.
114 Northfield Avenue
Edison, New Jersey 08837
USA

A QUANTUM BOOK

ISBN 0-7858-0165-0

QUMNRC

This book was produced by
Quantum Books Ltd
6 Blundell Street
London N7 9BH

Printed in Singapore by Star Standard Industries Pte Ltd

CONTENTS

INTRODUCTION

Once you have been taking riding lessons for a while, and you have read as much as you can about owning a horse, sit down and think it through carefully before making a final decision. Make a list of the pros and cons, and talk it over with your family and friends before you finally decide to go ahead. And once you have decided to buy a horse, you need to think about what sort of horse to get to suit what you want to do; and what will be the most convenient way for you to keep it.

Take a long, hard look at your riding ability and horse-management know-how. It is one thing to go riding once a week at the local riding school, but quite another to go out on your own and deal with a nervous horse in a tricky situation. If you know that your riding is quite good, but you doubt that you would be able, or willing, to handle a horse without some backup, take this into account when thinking about the type of horse you should look for and where you should keep it.

ABOVE: *One of the pleasures of owning your own horse lies in being able to ride out with friends.*

BELOW: *Your horse will need to be exercised every day whatever the weather conditions.*

LEFT: *As well as the main areas of show-jumping, dressage, and eventing, there are other competitive fields that you can enter, such as long-distance riding.*

Next there is the tack and equipment. This includes not just the saddle and bridle, but items like a halter, grooming kit, first-aid kit, and rugs. If you are keeping the horse yourself rather than at a livery stable, you will also need to include the costs of mucking out gear, feed bins, and hay nets.

Unless feed is included in the livery bill, you need to calculate the costs of concentrate feed, hay, and any supplements.

The horse will need to be shod, or at the least have its feet trimmed, every six to eight weeks, be wormed every two months, and have its teeth rasped twice a year. It will have to be insured and you will need to allow for extra funds to cover the veterinarian's bills in the case of an accident.

Do you have all the right clothing for yourself? Apart from the basics such as an approved hard hat or crash hat, jodhpurs, and boots, you may want your horse to compete. You need to take into account the cost of a show jacket or cross-country clothing, depending on what you want to do, together with safety equipment for your horse.

If you want to enter competitions you will need to take into account the cost of admission, that of transporting the horse, and of equipping it for traveling.

You should also include the cost of having lessons on your own horse. This will be invaluable in ironing out any problems you have and in preparing you for any shows or competitive events that you enter.

Finally, when buying the horse, remember to allow for the cost of having it checked over by a veterinarian, and for transporting it home.

The same applies to looking after a horse. It is one thing to know what to do when there are knowledgeable people around to support you, quite another when you are on your own.

Owning a horse is a big commitment, and it will take up a lot of your time. Horses need a routine that is kept to as closely as possible seven days a week. They need feeding and watering, mucking out and grooming. A single horse will thrive from your company, and a horse that is stabled will need even more attention than one which is out at grass.

When you are considering whether to buy a horse, you must take into account the cost of the whole enterprise. Firstly, there is the horse's accommodation, whether it is in a supervised yard where everything is done for you, down the road in a local farmer's field, or whether you are going to convert a building on your own land.

LEFT: *Horse sales are fascinating to visit, but the first-time buyer would be well-advised not to buy at them.*

For even an experienced rider, buying a horse can be an operation fraught with hazards. For example, the horse can turn out to be unsound, jittery, a rearer or a runaway, traffic-shy, bad in the stable, or difficult in the company of other horses or when left alone. It may, on the other hand, be a paragon of virtue, but simply not what you are looking for. Buying a horse is as highly personalized a procedure as choosing a wife or a husband.

There are many methods of buying a horse – riding magazines, for example, list horses for sale. But generally, if you are buying a horse for the first time, the most reliable course of action is to find a reputable dealer and rely on his judgment. This is far preferable to purchasing a horse at a sale. Sales are sometimes used to unload undesirable horses – the chronically sick, for instance, which have to be kept alive by drugs, or those which have serious vices. Of course, if a horse is warranted sound when it is sold and then proves not to be, then the purchaser can return the horse and get the money back. But it is

simpler and safer not to get into this situation in the first place.

Few reputable dealers will take advantage of someone who confesses their ignorance. The beginner should therefore admit his lack of experience and trust in the dealer's judgment, though an experienced friend is by far the greatest asset.

Points to watch for

There is a saying that a good horse should "fit into a box." This means that a classically conformed horse should, excluding its head and neck, be capable of fitting into a rectangle. A horse of this type is most likely to be, and remain, sound.

Good limbs are, of course, essential. The foreleg should give the overall impression of being "over," rather than "back" of the knee. Pay attention, too, to the horse's center of gravity – the part of the creature on which the greatest strain develops. Points before or behind the center are also liable to strain, but a well-conformed horse is far less at risk.

To assess the horse's personality, look it square in the eye; the character and intentions of a horse are fairly easy to read and interpret with a little experience. A bold but kind eye, generously proportioned, indicates a reliable, sympathetic temperament. Little piglike eyes, especially if the skull is convex between them and runs down to a Roman nose, are sure signs of an untrustworthy beast.

A Veterinarian's Role

Before any purchase is made, always have the horse examined by a veterinarian, who should be first told what the horse is required for. A hack, for instance, will not make an event horse. The examination should begin with the horse being "run up in hand," in order to check that the horse moves straight, and that it is sound. A sound horse can be heard to be going level and evenly, as well as seen to be doing so. An unsound horse will favor the lame leg, keeping it on the ground for as little time as possible. If very lame, it will nod its head as it drops its weight onto the sound leg.

The feet and limbs are then examined, the veterinarian being on the watch for any heat or swelling, exostoses (bony enlargements such as spavins, sidebones, or ringbones), and signs of muscular unsoundness, such as curbs, thoroughpins, or thickened tendons.

If all appears satisfactory, the horse's eyes are examined for cataract; it is then mounted and galloped to check its wind. This is to make sure that there are no hidden troubles with breathing or lungs – defects that are betrayed by a "roar" or a "whistle."

Horses with wind afflictions may

also have cardiac problems, for the effort of breathing in such cases naturally puts an added strain on the heart. For this reason, after the gallop, the heart is tested with a stethoscope.

The general condition of the horse is also examined and checks made for worms or other parasites. Finally, the veterinarian submits a report of his findings.

Trial before buying

It is sometimes possible to have a horse on trial for a limited period to see if horse and rider are compatible, though usually only if the dealer has a personal knowledge of the buyer. Horses are prone to all kinds of ailments and afflictions, and no dealer should be expected to entrust a horse to an inexperienced prospective purchaser.

A trial period is exceptionally valuable when buying a pony for a child. Here, the normal problems can be further compounded by the child's lack of strength, as well as, possibly, of experience. The safety of the child must be the first priority. Children have been killed when their ponies take fright and bolt – a particular hazard when riding on or near busy roads. It is therefore of the utmost importance only to buy from people with impeccable credentials. The outgrown family pony is ideal, but often difficult to find, as these animals are often passed onto the owner's friends and relatives. Ponies are also sold by their breeders, and breed societies will supply the names of studs.

For a first horse, do not make the mistake of buying one that is too young. A well-trained horse, that knows its job and is a willing and cooperative ride, is a much better buy than a young, inexperienced one. Two novices together is a bad combination; the horse is very likely to dominate its inexperienced rider.

Thus, a horse of four, five, or six years of age is not a beginner's ride.

At eight years, it is mature, and, provided that it is sound and healthy it should be useful and active until aged well over twenty. The more nervous the rider, the more docile the horse should be.

It is possible that, as a rider becomes more proficient, he or she will look for a horse with more quality. This is a natural and correct progression, but resist the temptation of buying a horse with too much "fire in its belly." This may well pull your arms nearly out of their sockets when, say, in company with other horses. Remember, too, that well-bred horses are far more expensive to keep than, say, a cob, for they usually have to be stabled in winter. A cob, on the other hand, can winter outdoors in a New Zealand rug quite happily, as long as it has access to a shelter and is given hay and one or two feeds a day.

BELOW: *Always have a horse thoroughly checked over by a veterinarian before you commit yourself to buying it.*

CHAPTER 1

CARE AND MANAGEMENT

KEEPING A HORSE AT GRASS

Looking after and caring for a horse or pony is perhaps the greatest responsibility any rider faces. Having learned to ride, many riders aim at eventually having a horse of their own. It is worth remembering, though, that looking after a horse unaided – especially if it is stabled – can be a full-time occupation. One answer is to board the horse at a livery stable. Another is to get someone to help out during the day. Most of the other factors involved, such as feeding, watering, exercising, and grooming, are mainly matters of common sense.

Horses can either be stabled, kept at grass or the two systems can be combined. This means that the horse can run free during the day and have the shelter of a stable by night – except in hot weather, when the procedure should be reversed. Whichever system is adopted is a matter of choice, practicality, and the type of horse concerned. Ponies, for example, are usually sturdier and more resilient to extremes of climate than horses, particularly thoroughbreds and partbreds. Some

thoroughbreds, for instance, should not be left outdoors over the winter. Nor can a horse being worked hard in, say, competitions be really kept fit enough, except by being cared for in a stable.

A combined system can also be adapted to suit the needs of a rider who is using his or her horse frequently, but cannot spare the time to keep it fully stabled. If the horse is being worked regularly in the spring or summer, say, it is a good idea to bring it into the stable first thing in the morning for the first extra feed that will be required. If the horse is to be ridden more than once that day, the same routine is followed as for a stabled horse until the afternoon, when the animal can be turned out for the night. If only one ride is possible, it can be turned out after the second feed, or, if it cannot be exercised at all, after the first.

Trees provide shelter from wind and sun

Stream provides fresh water

Gentle slope to stream

Easy access

Safe and secure fence

Undulating ground

LEFT: *Horses may roll to relax after being ridden, or just to deal with an irritating itch. Rolling can also be a symptom of colic, but generally it is simply a sign of pure enjoyment.*

The right field

Looking after a horse kept at grass is less time-consuming than looking after one kept in a stable. Among the advantages are the natural vitamins and the exercise the horse gets, but equal responsibility is still demanded from the owner. Statistics show that more accidents happen to horses left unattended in a field than those in a stable. They can kick each other, get tangled up in fences or gates, and quickly lose condition either through illness or just plain bad weather. Also, a horse should be visited every day, even if it is not being ridden. Horses are gregarious creatures – ideally, a horse should be kept in company with others – and they require affection.

The ideal field is large – between six and eight acres. It should be undulating, well-drained, securely fenced by a high-grown hedge reinforced by post-and-rail fencing, with a clump of trees at one end and a gravel-bedded stream to provide fresh water. But this situation is often hard to achieve. It is usually considered that about 1 to 1½ acres per pony is adequate, provided that the grass is kept in good condition. Because horses are "selective grazers" – that is, they pick and choose where and what they eat – a paddock can become "horse sick." Some places will be almost bare of grass, while others will be overgrown with the rank, coarse grasses the horses have found unpalatable. In addition, the ground will almost certainly be infested with parasites, the eggs of which horses pass in their dung. If action is not taken, the horses are sure to become infected with worms. These fall into two categories, of which nematodes (roundworms) are by far the most important and potentially destructive. Of these, the most dangerous are bloodworms (*Strongyles*), which, untreated, can lead to severe loss of condition. Even though the horse is well-fed, it looks thin and "poor," with a staring coat; in the worst cases, anemia may develop or indigestion, colic, and enteritis.

As far as an infected horse is concerned, the treatment is regular worming, but it is far better to tackle the problem at its source by making sure that the field is maintained properly. A large field should be subdivided so that one area can be rested while another is being grazed. Ideally, sheep or cattle should be introduced on the resting areas, as they will eat the tall grasses the horses have rejected. They will also help reduce worm infestation, as their digestive juices kill horse worms. Harrowing is also essential as it aerates the soil, encouraging new grass to grow, and also scatters the harmful dung. If this fails, the manure must be collected at least twice a week and transferred to a compost heap.

Mowing after grazing, together with the use of a balanced fertilizer, also helps to keep a field in good condition, but horses should not be returned to their grazing too soon after it has been treated in this way. If in doubt, allow three weeks.

Bots are another problem for field-kept horses, for which veterinary treatment is necessary.

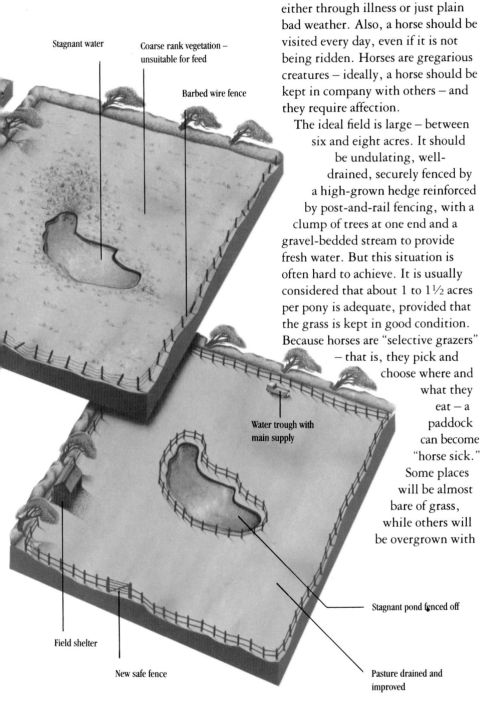

Stagnant water

Coarse rank vegetation – unsuitable for feed

Barbed wire fence

Water trough with main supply

Stagnant pond fenced off

Field shelter

New safe fence

Pasture drained and improved

When selecting a field for a horse, always aim for the ideal (top left), or, if the conditions are bad (center), improve them (above). A good field must be big enough to provide sufficient grazing – about one horse to 1¼ acres (0.5 hectare). Grass should be of good quality with no poisonous weeds. A supply of fresh water – preferably running – and some natural shelter are essential. The center field is thoroughly bad, with dangerous fencing, stunted, wind-swept trees, no gate, and a foul pond. What can be done with such unpromising material is shown (above). A new gate, sturdy fencing, water trough, field shelter, and improved pasture have transformed it into a suitable field for horses.

Food, water, and shelter

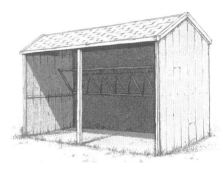

ABOVE: *A shelter is an essential addition to any field – even one with trees and hedges – as horses need to escape from wind and cold in winter. In summer it provides shade, coolness, and protection from insects. In cold or wet weather, hay can be conveniently fed in a rack or hay net within the shelter.*

BELOW: *Pastures vary according to area, but good grazing should include some of these grasses and weeds. Perennial ryegrass, timothy, and cocksfoot are the most nutritious and are readily sought out. Sainfoin, dandelion, and ribgrass are weeds with valuable mineral content. As horses are selective feeders and tend to overgraze, the various sections of the paddock need resting in turn through spring and summer to allow fresh growth. Grazing cattle or sheep on pasture ensures even grazing and will reduce worm infestation.*

All grassland is composed of a mixture of grasses and other plants. Some have little nutritional value, though the horse may well like them, but the three most important are perennial ryegrass *(Lolium perennae)*, cocksfoot *(Dactylis glomerata)* and timothy *(Phleum pratense)*. Some white clover *(Trifolium repens)* is useful, but beware of a heavily clovered pasture. This may prove too rich and lead to digestive problems.

Even if clover is not present, grass itself can cause problems. This is especially the case in the spring when excessive greed can lead a horse to put on too much weight, and sometimes to the painful disease called laminitis, or founder. Also, a horse or pony can only exist on grass alone for the summer months. By mid-fall, supplementary feeding becomes essential. Start with hay and then provide oats or beans, if required. The more refined the breed, the more extra feeding that will be necessary.

Water is another essential; field-kept horses must have easy access to a plentiful supply of fresh water. Remember that a horse drinks about 8 gal. (35 liters) a day. If the water supply is in the form of a stream, check that it can be reached by means of a gentle slope; if the banks are steep or muddy, it is safer to fence the stream off and provide a water trough instead. Similarly,

always fence off stagnant pools and ponds.

The most convenient form of trough is one connected to a mains water supply, controlled either by a faucet or automatic valve. Custom-made troughs are on the market, but cheaper alternatives are an old domestic cistern or bathtub. Remember to remove all sharp protrusions, such as faucets, and clean the inside thoroughly before putting it into use. If there is no piped water supply, use a hose or fill the trough with buckets.

Buckets on their own are insufficient. A horse can easily drink a whole bucket of water at one go, and, in any case, a bucket can all too easily be spilled. Daily checks of the water supply are vital, especially in winter, when ice may form and must be broken. A child's rubber ball left floating on the surface of a trough will help to keep the water ice-free, except when frosts are severe, when the ice must be broken daily.

Winter and summer also bring the problem of shelter. From a horse's point of view, the worst elements are wind, rain, and sun. Even if the field possesses a natural windbreak, an artificial shelter is a good addition. It need not be complicated – a three-sided shed the size of a large loose box is usually adequate. Make sure that the open side does not face the sun.

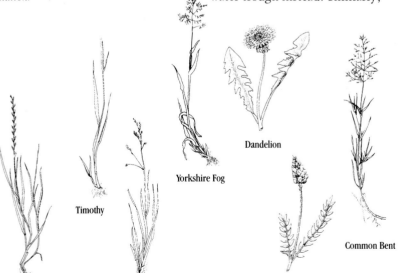

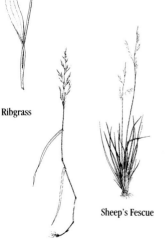

Dandelion

Ribgrass

Yorkshire Fog

Timothy

Common Bent

Sheep's Fescue

Perennial Rye Grass

Cocksfoot

Sainfoin

Meadow Fescue

Purple Moor Grass

Fencing and gates

BELOW: *Suitable types of fencing. From the left: post and rail, post and wire, rail and wire combined and dry stone wall. Check the tautness of wire fences regularly and inspect walls for damage after frosts.*

Sound and strong fencing is essential for safety. A fence must be high enough to prevent horses from jumping over it – 3ft 9in (1.3m) is the absolute minimum. Bars must also be fitted; two rails are usually adequate for containing horses, with the bottom one about 18in (45cm) from the ground. Small ponies, however, can wriggle through incredibly small gaps, so a third or even a fourth rail should be added for them. This type of fencing is known as post and rail.

Of all the types of fencing available, timber is the safest but most expensive. Hedges run a close second, but should be regularly checked, as a determined pony might otherwise well push his way through. Gaps can be reinforced with wood, but avoid filling a gap with wire. Concealed by a hedge in summer, it could be hard for a horse to see and so could lead to accidental injury. Stone walls are also attractive, but, they, too, will need regular checking, especially after a hard winter when frost may have loosened the mortar.

However, wire is perfectly adequate as fencing on its own, as long as the correct type of wire is used. Avoid barbed wire, or chicken and sheep wire. Use a plain heavy-gauge, galvanized wire instead. For safety and effectiveness, the strands must be stretched so that they are evenly taut and then stapled to the inside of the posts. Strong stretcher posts should be positioned at regular intervals. Check regularly for signs of weakness, such as loose posts, broken wires, or loose staples. If each strand of wire ends in an eye bolt attached to the end posts, the wire can be tightened from time to time.

Gates are another safety factor. The only criterion is that they should be easy for people to open and close, but that it should be impossible for the horse to do so. A five-barred farm gate, hung just clear of the ground so that it swings freely when unlatched, is ideal. It should be fitted with either a self-closing latch, or with a simple chain fitted with a snap lock and fastened to the latching post. Slip rails and hang-gates are cheaper alternatives.

ABOVE LEFT: *Self-filling trough with automatic valve in enclosed section and (right) trough with inlet pipe close-fitted and tap recessed beneath. Both are of safe design with no sharp edges or projections. Site a trough on well-drained ground to avoid churned mud and and away from the falling leaves of trees. Troughs should be emptied and cleaned regularly. If ice forms in the winter it should be broken daily.*

ABOVE: *Post-and-rail fencing made of good timber is the safest kind. It must be firm and strongly built. Trees and hedges provide a natural windbreak and shelter from the rain. They also shade the horse from the sun.*

KEEPING A HORSE STABLED

There are two main reasons for keeping a horse in a stable. The first is that the horse may by too well-bred to live outside in all weathers, without seriously losing condition. The second is the amount of work the rider requires the horse to do. If a horse is being ridden a great deal, it must be fit enough to cope with its rider's demands without showing signs of distress, such as excessive sweating and blowing. Such a degree of fitness takes time to achieve and can only be maintained in a stable.

The ideal stable is also often easier to provide than the ideal field. It should be roomy, warm, well-ventilated yet draft-free, easy to keep clean, have good drainage and be verminproof. It should face away from prevailing winds and have a pleasant outlook – preferably onto a stable yard or at least an area where something is often going on. The horse could be spending 22 hours a day in the stable and, unless there is something to holds its attention, it will probably become bored. This can lead, in turn, to the development of vices, such as weaving (rocking from side to side), box walking (a constant, restless wandering around the box), or crib-biting (gripping the manger or stable door with the teeth and drawing in a sharp breath). The first two vices may lead to loss of condition, the third to heaves.

Buying, renting, or building

Any stable, whether it is bought, leased, converted, or specially built, must conform to certain basic standards which must be followed. An architect can either design a stable to your individual specifications or one can be bought ready-made. This type of stable is usually delivered in sections and erected on a pre-prepared concrete base. But, before committing yourself, always check your plans out with the local authority concerned. They might have to grant planning permission, and will certainly have regulations governing such crucial health factors as drainage.

The choice of site is very important. As far as possible, it should be level and well-drained, with easy access to the electricity and water supplies. The stable itself should be situated with the doorway facing the sun and the general layout

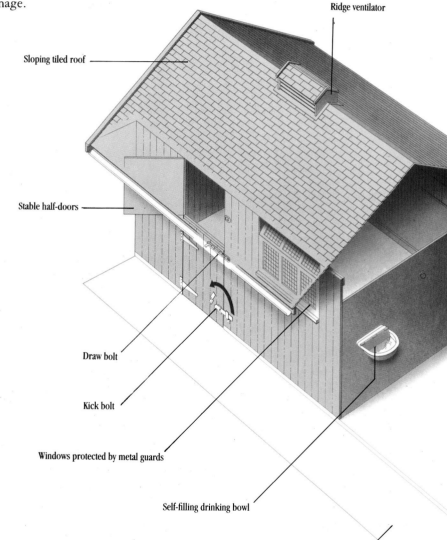

Ridge ventilator

Sloping tiled roof

Stable half-doors

Draw bolt

Kick bolt

Windows protected by metal guards

Self-filling drinking bowl

Concrete access path

Non-slip concrete floor

ABOVE: *Inside and outside views of two loose boxes planned with the comfort of the horse and ease of maintenance as the first considerations.*

should be planned to have all the essential elements – stable, feed room, tack room, and manure heap – conveniently close together.

The stable can either be a plain stall or a loose box; the latter is much more commonly used today, particularly in the UK and USA. The chief advantage of a stall is that it can be relatively small, which makes cleaning easier. But, as it is open at one end, the horse has to be kept tied up. The usual method is known as the rope and ball system,

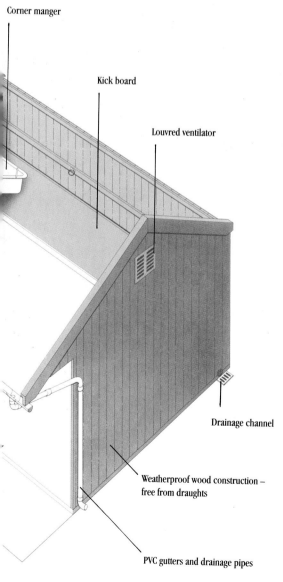

Corner manger

Kick board

Louvred ventilator

Drainage channel

Weatherproof wood construction – free from draughts

PVC gutters and drainage pipes

where the halter rope is passed through a ring on the manger and attached to a hardwood ball resting on the horse's bed. This helps to safeguard the horse against possible injury, while still allowing it some freedom of movement.

Most horse-owners prefer the loose box, as it allows the horse far more freedom to move around and so more comfort. Size here is all-important; cramming a 16hh hunter into a loose box built for a Shetland pony can only lead to trouble. As a rough guide, 12ft (3½m) square is probably the optimum size, rising to 13 ft (4m) for horses over 16hh. It is worth bearing in mind that a child's first pony, say, will be outgrown in time, so the bigger the box the better.

Boxes should be square rather than oblong, so that the horse can more easily determine the amount of room it has to lie down or to roll. The box must be big enough to minimize the risk of the horse being "cast" – rolling over and being trapped on its back by the legs striking the wall. In its struggles to get up, the horse may injure itself severely. The ceiling height should allow plenty of clearance for the horse's head; 10ft (3m) is the absolute minimum.

Brick and stone are both durable and attractive building materials, but solid concrete blocks, or timber may be cheaper. Both walls and roof should be insulated, which will keep the stable warmer when the weather is cold and cooler when it is hot.

The floor must be hard-wearing, nonabsorbent and slip-proof. A well-compacted concrete base is perfectly adequate, provided that it is made with a loam-free aggregate and treated with a proprietary nonslip coating after laying.

Alternatively, roughen the surface with a scraper before the concrete sets. Make sure that the floor slopes slightly – a slope of

about one in sixty from front to rear is ideal – so that urine can drain away easily. An alternative is to cut a narrow gully along one inside wall leading to a channel in the wall and so to an outside drain. The channel should be fitted with a trap to stop rats from getting in, and cleared of dirt and debris daily.

The usual type of stable door is made in two halves, the top half being kept open for ventilation. This should be planned to make sure that the horse gets plenty of fresh air but no drafts, as these can lead to it catching colds and chills. The best position for a window is high on the wall opposite the door so that sufficient cross-ventilation can be provided. Make sure it is fitted with shatterproof glass, and covered with an iron grille. Otherwise, vents can be built in the roof to allow stale air to escape. They should be protected by cowls.

Doors must be wide and high enough for a horse to pass through without the risk of injury; 4ft (1.5m) is the minimum width, 7½ft (2.25m) the minimum height. Make sure that the door opens outward so that access is easy and that strong bolts are fitted to both halves of the door. On the lower door, two bolts are necessary – an ordinary sliding bolt at the top and a kick bolt, operated by the foot, at the bottom. The top half needs only one bolt. Remember that the material used must be strong enough to withstand the kicking of a restless horse. Inside the stable, kicking boards will help with this problem.

Electricity is the only adequate means of lighting. The light itself should be protected by safety glass or an iron grille and all wiring should be housed in galvanized conduits beyond the horse's reach. Switches must be waterproof, properly insulated, and, whenever possible, fitted outside the stable.

Fixtures and fittings

The basic rule to follow is the fewer fittings, the better, to minimize the risk of possible injury. The only essential is a means of tying the horse up. Normally, this consists of two rings, fixed to bolts which pass right through the stable wall. One ring should be at waist height and the other at head height. All other fittings and fixtures are a matter of individual preference.

Fixed mangers positioned at breast level and secured either along a wall or in a corner of the loose box are found in many stables. They should be fitted with lift-out bowls to facilitate cleaning and have well-rounded corners. The space beneath should be boxed up to prevent the horse from injuring itself on the manger's rim – this space makes a good storage place for a grooming kit. However, a container on the floor, which is heavy enough not to be spilled and which can be removed as soon as the horse has finished its feed, is adequate.

Fitted hay racks are found in some stables, but they are not really advisable. They force the horse to eat with its head held unnaturally high and hayseeds may fall into its eyes. The best way of feeding hay is to use a hay net. It is also the least wasteful, as hay nets permit accurate weighing. The net should always be hung clear of the ground and be fastened with a quick-release knot to one of the tying-up rings.

Water is as essential to the horse in the stable as for a horse in the field. Automatic watering bowls are one way of providing a constant supply – but never position them too close to the hay net and manger, or they may get blocked by surplus food. Buckets are satisfactory, provided, that the bucket is heavy enough not to be accidentally upset. Using a bucket means that it is possible to control the amount of water the horse drinks – important after exercise, for instance, when a "heated" horse must not drink too much – and also to check how much it is drinking more easily. This is especially useful in cases of suspected illness.

BELOW: *The daily routine for a fully stabled horse, showing the order of work and the times at which different tasks are carried out. The feeding schedule will vary according to the size and workload of individual horses. Many owners prefer the less time-consuming combined system, in which horses spend part of the day out in the field.*

Stable routine

The daily program for looking after a stabled horse takes up a great deal of time (see below). All the stages have to be carried out, though some, such as the number of feeds, will vary from case to case. Skimping will only lead to problems later.

The only way of short-circuiting this routine is to adopt the combined system of care. This has considerable advantages in time and labor, but is not suited to all horses, especially those being worked hard. Otherwise, board or livery is the only alternative. Some riding schools offer what is termed half-livery; this means that the horse gets free board in exchange for use as a hack. The risk is that the horse may be roughly treated by inexperienced riders even in a supervised lesson. Fully livery, however, is extremely expensive. In either case, always check that the stable you choose is officially approved by a recognized riding authority.

The principal areas of a horse-owner's day, however, are not as complex as they seem. They can be broken down into various tasks, all of which are relatively simple to carry out.

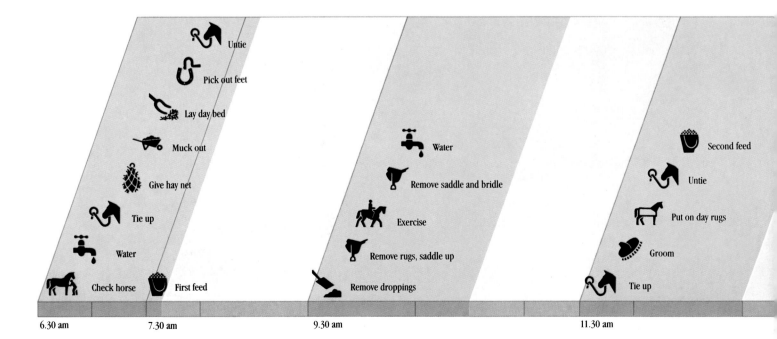

Untie
Pick out feet
Lay day bed
Muck out
Give hay net
Tie up
Water
Check horse   First feed

Water
Remove saddle and bridle
Exercise
Remove rugs, saddle up
Remove droppings

Second feed
Untie
Put on day rugs
Groom
Tie up

6.30 am   7.30 am   9.30 am   11.30 am

Bedding down and mucking out

The purpose of bedding is to give the horse something comfortable to lie on, insulate the box, absorb moisture and prevent the horse's legs jarring on the hard stable floor. It must be kept clean – hence the daily task of mucking out. This is usually done first thing in the morning, and, with practice, can be carried out quite quickly.

Straw is the best possible bedding material, though other kinds can be substituted. Wheat straw is excellent, because it is absorbent and lasts well. Barley straw may contain awns, which can irritate the horse's skin. Oat straw should be avoided, because horses tend to eat it and it tends to become saturated.

Of the substitutes, peat makes a soft, well-insulated bed; it is also the least inflammable of all bedding materials. However, it is heavy to work. Damp patches and droppings must be removed at once, and replaced with fresh peat when necessary. The whole bed requires forking and raking every day, as the material can cause foot problems if it becomes damp and compressed.

1 Mucking out is the first job to be done each morning in the stable. Soiled straw and dung are separated from the cleaner portions of the night bedding by tossing with a fork. The cleaner straw is then heaped at the back of the stall to be used again.

2 The soiled straw and droppings are put into a wheelbarrow for removal to the manure heap. In fine weather much of the night bedding can be carried outside to air in the sun. This will freshen it up, restore its springiness, and make it last longer.

3 When the bulk of the soiled straw has been removed and the cleaner straw reserved, the floor should be swept clean of remaining dirt. It should be left bare to dry and air for a while. The clean straw is then spread as a soft floor covering for the day.

4 The soiled straw and dung are tossed onto the manure heap. Take care to throw the muck right onto the top of the heap, as a neatly built heap decomposes more efficiently. Beat the heap down with a shovel after each load to keep it firm and dense.

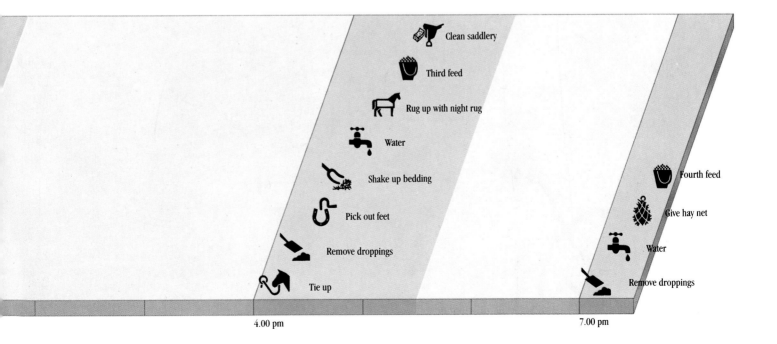

Clean saddlery

Third feed

Rug up with night rug

Water

Shake up bedding

Pick out feet

Remove droppings

Tie up

Fourth feed

Give hay net

Water

Remove droppings

4.00 pm

7.00 pm

Wood shavings and sawdust are usually inexpensive but can be difficult to get rid of. Both need to be checked carefully to see that they do not contain nails, screws, paint, oil, or other foreign matter. Wood shavings can be used alone, but note that they can cause foot problems if they become damp and compressed. Sawdust is best used in combination with other materials.

There are two types of bed – the day bed and the night bed. The first is a thin layer of bedding laid on the floor for use during the day; the second is thicker and more comfortable for use at night. With materials such as peat or wood shavings, laying the bed is very simple. Just empty the contents of the sack on the floor and rake them level, building up the material slightly higher around the walls to minimize drafts.

Laying a straw bed requires slightly more skill. As the straw will be compressed in the bale, it has to be shaken so that the stalks separate, and laid so that the finished bed is aerated, springy, and free from lumps. A pitchfork is best for the purpose.

Some owners prefer the deep litter method of bedding, where fresh straw is added to the existing bed every day, removing only droppings and sodden straw beforehand. After a time, the bed becomes as much as 2ft (60cm) deep, well-compressed below and soft and resilient on the surface. At the end of a month, the whole bed is removed and restarted. This method should be used only in loose boxes with first-rate drainage. In addition, the feet must be picked out regularly; otherwise there is a major risk of disease.

BELOW: *Paper and straw beds at the Royal Mews, Windsor.*

LAYING A STRAW NIGHT BED

1 Clean straw saved from the day bed is tossed and shaken well with a pitchfork before being spread evenly over the floor as a foundation.

2 New straw is taken from the compressed bale and shaken well to free the stalks and make the bed springy. The floor must be thickly and evenly covered to encourage the horse to lie down.

3 The straw is banked up higher and more thickly around the sides of the box. This reduces drafts, keeps the horse warmer, and gives the animal extra protection from injury during the night.

TYPES OF BEDDING

ABOVE: *Wheat straw makes ideal bedding. It is warm, comfortable, easy to handle, and absorbent.*

ABOVE: *Wood shavings make cheaper bedding and are often laid on a base of sawdust to reduce dampness. Droppings have to be removed frequently.*

ABOVE: *Paper bedding can be used in cases where horses tend to eat the straw, but it needs changing more frequently.*

FEEDING

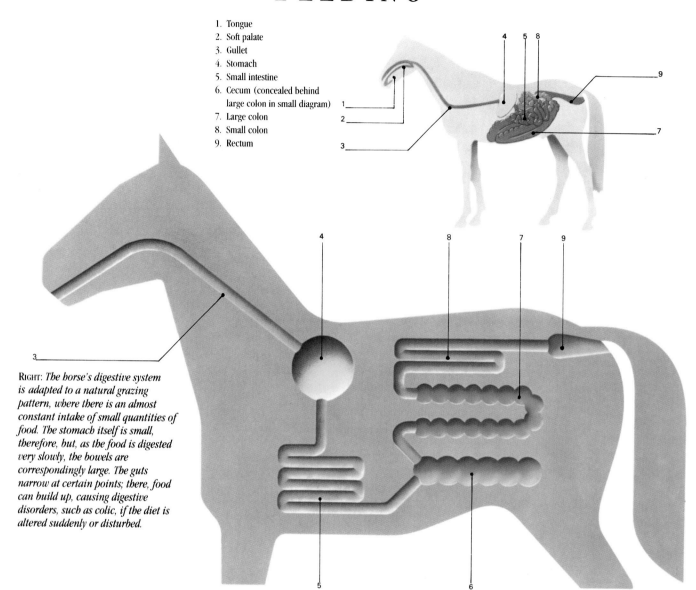

1. Tongue
2. Soft palate
3. Gullet
4. Stomach
5. Small intestine
6. Cecum (concealed behind large colon in small diagram)
7. Large colon
8. Small colon
9. Rectum

RIGHT: *The horse's digestive system is adapted to a natural grazing pattern, where there is an almost constant intake of small quantities of food. The stomach itself is small, therefore, but, as the food is digested very slowly, the bowels are correspondingly large. The guts narrow at certain points; there, food can build up, causing digestive disorders, such as colic, if the diet is altered suddenly or disturbed.*

Heredity has given the horse a very small stomach for its size and the food it eats takes up to 48 hours to pass through the digestive system. This system is in itself complex. It depends not only on the right amounts of food at the correct time for smoothness of operation, but also on an adequate supply of water and plenty of exercise. In the wild, horses drink twice a day, usually at dawn and dusk. In between, their day is divided into periods of grazing, rest and exercise. Field-kept horses can duplicate this pattern to some extent, but stabled horses cannot do so.

It is essential to follow a basic set of feeding rules. Otherwise the horse's sensitive digestion might be upset, encouraging the risks of indigestion, impaction, formation of gas in the stomach, or sudden colic attacks.

The basic rules are to feed little and often, with plenty of bulk food – grass or hay – and according to the work you expect the horse to do. Make no sudden change in the type of food, or in the routine of feeding, once the diet and time has been established. Always water the horses before feeding, so that indigested food is not washed out of the stomach. Never work a horse hard immediately after feeding or if its stomach is full of grass. Let it digest for 1¼ hours or so, otherwise the full stomach will impair breathing. Similarly, never feed a horse immediately after hard work, when it will be "heated."

The staple diet of the horse is grass, or, in the case of a stabled horse, hay. The best type is seed hay, usually a mixture of ryegrass and clover, which is specially grown as part of a crop-rotation program. Meadow fescue, also commonly used, comes from permanent pasture and so can vary in quality. The best way of judging this is by appearance, smell, and age. Hay

LEFT: *Basic concentrated foods are an essential part of the diet for horses in hard, regular work. 1. Bran is rich in protein, vitamin B, and salt. Fed as a mash, or slightly damp, with oats, it has a laxative effect. 2. Oats are a balanced, nutritious and easily digested food, high in energy-giving carbohydrate, vitamin B, and muscle-building protein. They are fed whole, bruised or crushed. 3. Sugar beet cubes provide bulk for horses in slow work. They must be soaked before use, or will swell in the stomach and cause colic. There is also a great danger of the horse choking. 4. Corn, fed flaked for easy digestion, is energizing, but low in protein and minerals. It contains vitamin A. 5. Barley, unsuitable for horses in long, fast work, is fed boiled, as a general conditioner; it should be crushed if fed raw. It contains vitamin B.*

should smell sweet, be slightly greenish in color and at least six months old. Blackened, moldy, or wet hay should never be used as fodder.

Of the other types of hay, clover is too rich to be fed to a horse on its own, and the same rule applies to alfalfa, or lucerne, common in the USA and Canada. Alfalfa is extremely rich in protein, so feed small quantities until you can judge how much is needed.

Concentrates for work

Ponies and horses in regular, hard work need additional food to keep them in a fit, hard-muscled condition. In other words, they need energy rather than fatness. This is provided by the feeding of concentrated foodstuffs, usually known as "short" or "hard" feeds. Of these, the best is oats, which can be bruised, crushed, or rolled to aid digestion. Manufactured horse cubes or pellets are a useful alternative.

Oats have no equal as a natural high protein, energy-giving food and are an essential part of the diet for all horses in work. Good quality oats are plump and short, and pale gold, silver gray, or dark chocolate in color. They should have a hard, dry feel and no sour smell. Do not feed too much, or a horse may speedily become unmanageable. This caution applies particularly to children's ponies, which are often better off without oats at all.

Cubes and pellets are manufactured from various grains and also usually contain some grass meal, sweeteners such as molasses or treacle, extra vitamins and minerals. Their nutritional value is about two-thirds that of oats, but they are less heating and so ideal for ponies. Their chief advantage is that they provide a balanced diet on their own, as they do not have to be mixed with other foodstuffs. However, they are expensive.

Other grains can be used in addition or as alternatives to oats, but they are all of lesser quality. Flaked corn is used in many parts of the world as a staple feed. It is high in energy value, but low in protein and mineral content. Like oats, it can be heating for ponies and is usually fed to animals in slow, regular work, such as riding-school hacks. Boiled barley helps to fatten up a horse or pony in poor condition and is a useful addition to the diet of a stale, or overworked, horse. Beans, too, are nutritious, but again, because of their heating effect, they should be fed sparingly, either whole, split or boiled.

Other useful foods

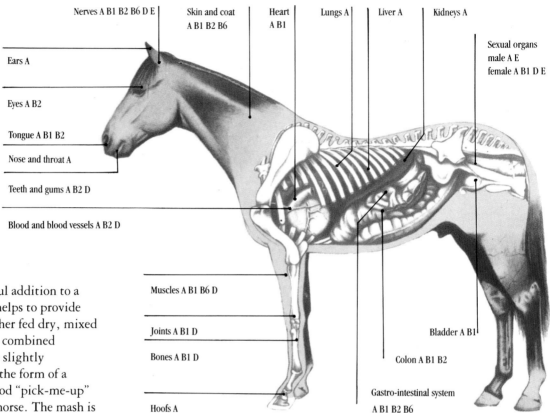

Nerves A B1 B2 B6 D E

Skin and coat A B1 B2 B6

Heart A B1

Lungs A

Liver A

Kidneys A

Sexual organs male A E female A B1 D E

Ears A

Eyes A B2

Tongue A B1 B2

Nose and throat A

Teeth and gums A B2 D

Blood and blood vessels A B2 D

Muscles A B1 B6 D

Joints A B1 D

Bones A B1 D

Bladder A B1

Colon A B1 B2

Gastro-intestinal system A B1 B2 B6

Hoofs A

Bran makes a useful addition to a horse's diet, as it helps to provide roughage. It is either fed dry, mixed up with oats – the combined mixture should be slightly dampened – or in the form of a mash. This is a good "pick-me-up" for a tired or sick horse. The mash is made by mixing ⅔ of a bucket of bran with ⅓ of boiling water and is fed to the horse as soon as it is cool enough to eat. Always remove any remains, as the mash can quickly go rancid. Oatmeal gruel is an alternative. This is made by pouring boiling water onto oats and leaving it to cool. Use enough water to make the gruel thin enough in consistency for the horse to drink.

Linseed, prepared as a jelly, mash, or tea, is fed to horses in winter to improve condition and to give gloss to the coat. It must be soaked, then well-cooked to kill the poisonous enzyme present in the raw plant. Allow the mix to cool before giving it to the horse. Dried sugar beet is another good conditioner, because of its high energy content. Most horses like it because of its sweetness. It must be always soaked in water overnight before it is added to a feed. If fed dry, the beet is likely to cause severe colic, as it swells dramatically when wet.

Roots, such as carrots, turnips, and rutabaga, also help condition

ABOVE: *Proper feeding with the correct balance of vitamins is essential for health. This diagram shows how particular vitamins work throughout the system and what effects they*

have. Any deficiency of these vitamins, A, B1, B2, B6, D and E, in the horse's diet, will eventually lead to debility and general loss of condition.

ABOVE: *Foodstuffs should be kept in separate bins and only mixed at feedtime. A scoop is used to measure out each ration. Where several horses are kept a checklist for feeding should be pinned up near the bins so that each receives its appropriate diet. The feed can be dampened slightly with water before being mixed by hand in a bucket and then fed to the horse. Grain keeps fresh and dry in galvanized bins. The lids should be heavy enough to prevent a horse from raising them.*

and are also of particular value to fussy feeders. Always wash the roots first and then slice into finger-shaped pieces. Small, round slices may cause a pony to choke.

Molasses or black treacle can be mixed with food to encourage a finicky feeder. In any case, all feeds ideally should contain about 1lb (450g) chaff – chopped hay. Chaff has practically no nutritional content, but it does ensure that the horse chews its food properly, thus helping to minimize the risk of indigestion. It also acts as an abrasive on teeth. Finally, a salt or mineral lick – left in the manger – is essential for all stabled horses. Field-kept animals usually ingest an adequate amount of salt during grazing, but a lick is also a good safeguard.

RIGHT: *A separate food store is essential. It should be clean, dry, and near to the water supply. Foodstuffs kept in the stable can easily become spoiled or contaminated. The horse is a fastidious feeder, and musty or dusty food, as well as being unappetizing, may be harmful. This simple food store provides a clean, secure, and compact area where foodstuffs can be measured out and mixed. Scales are also useful to check the weight of filled hay nets periodically.*

Vitamins and minerals

An adequate supply of vitamins and minerals is vital in addition to the required amounts of carbohydrates, proteins, and fats. Vitamins A, B1, B2, B6, D, and E are all essential; otherwise the horse's resistance to disease will certainly be lessened, and actual disease is likely to result. Normally, good-class hay and grass, bran, and carrots will contain most of the vitamins a horse needs; oats and barley, flaked corn, and sugar beet pulp are also all useful. Vitamin D can only be artificially administered through cod-liver oil, or left to the action of sunlight on the natural oil in the coat.

The absence of a sufficient supply of minerals can be even more serious than a lack of vitamins, especially in the case of a young horse. The essential minerals required are: calcium and phosphorus, for the formation of healthy teeth and bones; sodium, sodium chloride (salt), and potassium, for regulation of the amount of body fluids; iron and copper, vital for the formation of hemoglobin in the blood to prevent anemia; while magnesium, manganese, cobalt, zinc, and iodine are all necessary. Magnesium aids skeletal and muscular development; manganese is needed for both the bone structure and for reproduction; zinc and cobalt stimulate growth; while iodine is particularly important to control the thyroid gland.

However, of all these minerals, the most important is salt. This is why it is vital to provide a horse with a salt lick in either the stable or field.

As with vitamins, the chief source of these minerals is grass or hay, together with the other foods mentioned above. However, if the horse needs extra vitamins and minerals, always take the advice of a veterinarian first – an excess of vitamins or minerals can be as dangerous as an underdose. There are many suitable proprietary products on the market. These usually come in the form of liquid, powders, and pellets, designed to be mixed in with other food.

Signs of lack of vitamins are usually seen on the skin and coat; examination of the teeth, gums, and eyes can also give warning of possible deficiency. But, with sensible and controlled feeding, the problem should not arise.

Quantities to feed

There is no rigid guide as to the exact amounts of food a horse should be fed; much depends on the type and size of horse and the work it is expected to do. However, as far as a stabled horse is concerned, the amount should certainly not be less than the horse would eat if it was grazing freely.

If the horse concerned was 15.2hh, say, it would eat 26½ lb (12 kg) of grass a day. Bigger horses require an extra 2 lb (1 kg) for every extra 2in of height; smaller ones need 2 lb (1 kg) less.

With this basic total established, it is possible to plan a feeding program, varying the amounts of bulk and concentrated food according to the demands being made upon the horse. Taking as an example a lightweight 15.2hh horse that is being hunted, say, three days a fortnight in addition to other regular work, the emphasis will be on an almost equal balance between concentrated food and hay or grass. The horse should be getting about 14 lb (6.3 kg) of concentrates a day to some 15 lb (6.8 kg) of hay. If, however, the horse is being lightly worked – or not worked at all – the amount of hay will rise and the quantities of concentrated food will diminish.

Remember, too, that most horses feed much better at night, so it is important that the highest proportion of food be given in the final feed of the day. If the horse is being given three feeds a day, for example, the proportions are 10 percent in the morning, 30 percent at midday and 60 percent at night.

The best guide of all is simple observation. If a horse is too fat, it will need its rations reduced; if too thin, it will need building up.

GROOMING

The chief point of grooming is to keep the horse clean, massage the skin, and tone up the muscles. Field-kept horses need less grooming than stabled horses, particularly in winter, but some must nevertheless be carried out.

A good grooming kit is essential. This should consist of a dandy brush, to remove mud and dried sweat marks; a body brush, a soft, short-bristled brush for the head, body, legs, mane, and tail; a rubber currycomb, used to remove thickly caked mud or matted hair, and a metal one, for cleaning the body brush; a water brush, used damp on the mane, tail, and hooves; a hoof pick; a stable rubber, used to give a final polish to the coat; and some foam-rubber sponges, for cleaning the eyes, nostrils, muzzle, and dock.

Where more than one horse is kept, each animal should have its own grooming kit, kept together in a box or bag and clearly marked. This helps to prevent the risk of infection in cases of illness.

Grooming falls into three stages, each of which is carried out at a different time of the day. The first of these is quartering, normally done first thing in the morning before exercise. Tie up the horse. Then, pick the feet and, next, clean the eyes, muzzle, and dock with a damp sponge. If worn, rugs should be unbuckled and folded back and the head, neck, chest and forelegs cleaned with a body brush. Replace the rugs and repeat the process on the rear part of the body. Remove any stable stains with a water brush. Finish by brushing the mane and tail thoroughly with the body brush.

Strapping is the name given to the thorough grooming which follows exercise, when the horse has cooled down. Once again, tie the horse up and pick its feet. Follow by using the dandy brush to remove all traces of dirt, mud, and sweat, paying particular attention to marks left by the girth and saddle and on the legs. Work from ears to tail, first on the near side and then on the off. Take care to use the brush lightly to avoid irritating the skin.

Next, comes the body brush. This must be used firmly for full effect. Start with the mane, pushing it to the wrong side to remove scurf from the roots. Brush the forelock. Then, start on the body, working from head to tail and grooming the near side first, as before. Work with a circular motion, finishing in the direction of the hairs, and flick the brush outward at the end of each stroke to push dust away from the body. At intervals, clean the brush with the currycomb, which is held in the other hand. It can be emptied

LEFT: *The grooming kit. Ideally every horse should have its own, to reduce the chance of any infection being passed from one to another. Keep the kit in a wire basket or bag so that no item is mislaid. Clean the equipment from time to time with a mild disinfectant.*
TOP ROW: *1. Sponges, one for cleaning eyes, lips, and nostrils, the second for cleaning the dock. 2. Water brush (soft), for laying mane and tail and washing feet. 3. Mane combs are used when mane or tail is braided, trimmed, or pulled. 4. Can of hoof oil and brush, used to improve appearance of hoof and to treat brittle feet. 5. Metal currycomb, for cleaning dirt from body brush (never used on the horse). 6. Body brush (soft) to remove dust and scurf.*
BOTTOM ROW: *7. Rubber currycomb, removes dirt from body brush; can also be used in place of dandy brush. 8. Hoof pick, for taking dirt and stones from the feet. 9. Dandy brush (hard) to remove dried mud and sweat. 10. Stable rubber used for final polishing of coat. 11. Sweat scraper, to remove water and sweat from coat.*

GROOMING

by tapping on the floor occasionally.

Brush the head, remembering that this is one of the most sensitive areas of the horse. So use the brush firmly, but gently, and take particular care when grooming around the eyes, ears, and nostrils. Finally, brush the tail – a few hairs at a time – so that every tangle is removed.

The next stage is wisping, which helps to tone up the muscles and also stimulates the circulation. A wisp is a bundle of soft hay, twisted up to form a rope. Slightly dampen it, and use vigorously on the neck, shoulders, quarters, and thighs, concentrating on the muscular areas. Bang the wisp down hard on these, sliding it off with, not against, the coat. Take care to avoid bony areas and the tender region of the loins.

Sponge the eyes, lips, and muzzle and nostrils. Then, with a second sponge to minimize the risk of possible infection, wash around the dock and under the tail. Lift the tail as high as possible, so the entire region can be adequately cleaned. "Lay" the mane with the water brush. Then brush the outside of the feet, taking care not to get water into the hollow of the heel. When the hooves are dry, brush hoof oil over the outside of each hoof as high as the coronet.

Finally, work over the horse with the stable rubber for a final polish. The object is to remove the last traces of dust. Fold the rubber into a flat bundle, dampen it slightly, and then go over the coat, working in the direction of the lay of the hair.

Strapping takes from between half to three-quarters of an hour with practice. It will normally take a novice slightly longer, mainly because of the unaccustomed strain it imposes on the groom's muscles. "Setting fair" – the last grooming of the day – takes far less time. Simply brush the horse lightly with the body brush, wisp and then put on the night rug if one is normally worn.

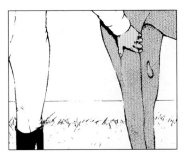

1 To pick up a horse's foot, stand facing its tail. Warn it first by sliding a hand down from its shoulder to its fetlock. This can also encourage the horse to move its weight over to the other legs.

2 Working from the frog to the toe end concentrating on the edges first, use the point of the hoof pick to prize out any foreign objects lodged in the foot. Take care not to push the point into the frog.

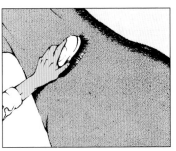

3 The dandy brush is used to remove heavy dirt, caked mud, and sweat stains, particularly from the saddle region, belly, fetlocks, and pasterns. As it is fairly harsh it should not be used on the more tender areas.

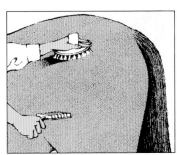

4 A body brush has short, dense bristles designed to penetrate and clean the coat. It should be applied with some pressure, in firm, circular movements. After a few strokes clear it of dust with a currycomb.

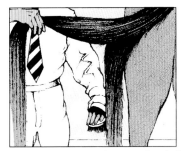

5 The body brush is also used to groom the tail. This should be brushed a few hairs at a time, starting with the undermost ones. Remove all mud and tangles. Finally, the whole tail should be brushed into shape from the top.

6 Wring out a soft sponge in warm water and sponge the eyes first. Carefully sponge around the eyelids. Wring out the sponge and wipe over the muzzle, lips, and nostrils. A separate sponge should be used for the dock area.

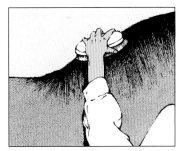

7 The water brush is used to "lay" the mane. The tip of the brush is dipped in a bucket of water and thoroughly shaken before it is applied. Keeping the brush flat, make firm, downward strokes from the roots.

8 As a final touch to the grooming, go over the whole coat with the stable rubber to remove any trace of dust. This cloth is used slightly damp and folded into a flat bundle. Work along the lie of the hair.

Checking for injury

The purpose of checking a horse for injury is obvious. You must make sure the horse has not cut, scratched, or bruised himself and check that he appears to be in general good health. Run your hand over his coat to feel for lumps and bumps, paying particular attention to his legs. Bruises will manifest themselves by obviously tender or hot areas, or by swellings. Pick out each foot in turn to make sure no stones have lodged in the hoof, or that the sole of the hoof has not been bruised or pricked by stepping on something sharp.

Any small cuts you may find will generally need very little treatment. Wash them well, preferably using running water from a hose, to clean them and then check the extent of the injury. Dress them with antibiotic powder, which you can get from your veterinarian – you should always keep a supply handy.

If a horse has sustained a more serious wound – from some broken glass someone has tossed into the field, for instance – do not hesitate to call the veterinarian. Proper medical attention in such instances not only speeds recovery, but it often reduces the possibility of the horse being marked by an unsightly scar later.

ABOVE: *A habit you should get into is checking your horse for injuries. This is especially important if your horse is at grass, as you won't be able to watch him. Whether done while he is still in the field, or in the stable, a thorough check should be made of his entire body, especially the legs and hooves as these are the most prone to cuts, sprains, and infection. Having tethered the horse, move your hands up and down each of the legs, feeling for any swelling, and watching to see if he reacts with pain.*

RIGHT: *If a minor injury is found, it should first be thoroughly washed with warm water and then sprayed or wiped with antiseptic. Never leave a cut unattended.*

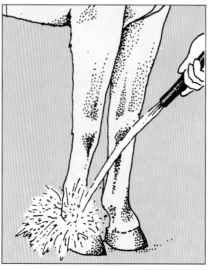

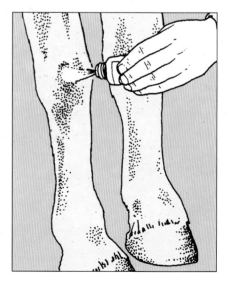

Shampooing a horse

You should always give your horse a thorough grooming or strapping the day before a show. If he has any white socks or stockings, these should be shampooed. The tail, too, will look smarter if it is washed.

Hard, yellow soap available from saddlers should be used when shampooing any part of a horse; never use soft, cosmetic-type soap. To wash the legs, wet them first with tepid water, then rub them with soap to produce a lather. Scrub very gently with plenty of clean water, making sure all vestiges of soap have been removed. Rub the legs as dry as possible with a stable rubber, leave them to dry completely before bandaging them. If you wash your horse's legs frequently, grease the heels periodically to reduce the risk of the skin cracking due to excessive dampness.

To shampoo the tail, first brush it in the usual way to remove any mud and tangles. Then, wet it thoroughly and soap it well. Rub the hairs together with your hands and rinse the tail thoroughly. Run your hands down the length of the tail several times in order to remove as much water as possible. Then, standing by the side of the hindquarters, hold the tail by the bottom of the dock and swing it around in a circle. This helps to dislodge any remaining water in the same way as a dog does when it shakes itself after a bath or swim.

Brush the tail in the normal way, using the body brush and put on a tail bandage (see page 63) which helps to keep the tail clean and makes sure the top hairs dry flat. The bandage should not be left on overnight, as it can constrict the circulation in the dock. Thus, you have washed the tail the day before a show. Remove the tail bandage when you bed down the horse and put on another after brushing the tail in the morning.

If a horse's coat is very stained, it is possible to shampoo him all over, but this should only be done if it is absolutely necessary. Only shampoo a horse's coat if the day is warm, dry, and preferably sunny. Use the same sort of soap described above, with plenty of warm water, and, having soaped and scrubbed the entire coat, rinse it very thoroughly. Remove the surplus water by pulling a sweat scraper across the coat. Then go over the entire body with a dry sponge to "mop" as much remaining water as possible. After this, rub it dry, then walk the horse around until the coat has dried completely. Finally, brush the coat in the normal way with the body brush.

Clipping

Horses are clipped to maintain comfort and, less importantly, for smartness. Removing all or part of the coat by clipping prevents heavy sweating during exercise in winter and therefore lessens the risk of a horse catching a chill. It also enables the horse to dry off more quickly.

There are various types of clip; choice should depend on what the horse is expected to do, and how much it sweats doing it. Remember that a clipped horse will need to wear rugs for warmth during cold weather, whether it is kept in a stable or a field.

CLIPPING

Choosing to clip your horse's coat in winter allows him to work harder and keep him looking smart. The types of clips vary considerably. The trace clip removes just a line of hair from beneath the neck, belly, and tops of legs. The blanket clip is the same as the trace clip but with all hair removed from the neck. The hunter-clipped horse must be kept inside in winter or turned out with a warm New Zealand rug.

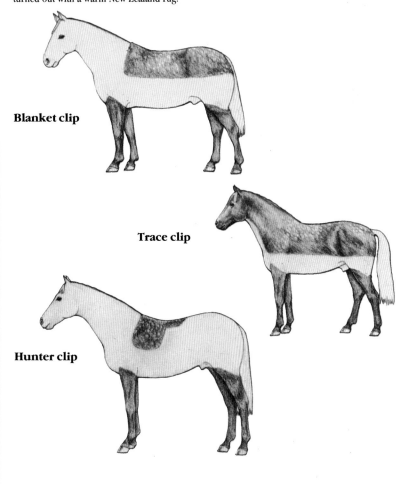

Blanket clip

Trace clip

Hunter clip

SHOES AND SHOEING

H orse maintenance includes not only the routines of feeding and cleaning, but also very important areas such as shoeing and illness for which you should establish a good relationship with a qualified farrier and veterinarian.

Horses that are being worked to any extent – whether they are being ridden or used to pull a vehicle – must have their feet shod with metal shoes. Riding an unshod horse soon wears away the hard, insensitive horn of the foot and exposes the more sensitive areas. If this happens, these areas become sore and the horse becomes lame within a very short time.

The shoeing of horses is a highly skilled and specialist task, undertaken by a trained craftsman known as a farrier. If the shoes do not fit, or if they pinch the feet in any way, then obviously the horse will not be able to give its best.

To understand how the farrier goes about his task, it is necessary to have some knowledge of the structure of the horse's foot.

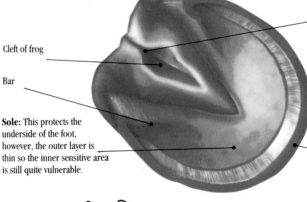

ABOVE: *The farrier's skill with his tools will mainly determine the fit of the horse's shoe and, consequently, his performance.*

TYPES OF SHOES

RIGHT: *The care of the horse's feet is probably the most important part of horsemanship and every rider should know the parts and functions of the horse's hoof.*

BELOW: *Several types of shoes including a hunter shoe, a grass shoe, a corn shoe, a leather shoe, and a T-shaped shoe for horses with contracted heels or corns.*

Cleft of frog

Bar

Sole: This protects the underside of the foot, however, the outer layer is thin so the inner sensitive area is still quite vulnerable.

Frog: The frog is V-shaped and leathery, and provides the foot with a natural shock absorber and non-slip device. The farrier never pares back the frog and it needs daily attention to keep it clean and healthy.

Wall: The wall, like the human fingernail, is insensitive and always growing. Because of the latter, the shoes must be removed regularly and the hoof pared into shape.

Hot and cold shoeing

There are two types of shoeing – hot and cold. At one time, the former was virtually the only method practiced, but nowadays it has been largely superseded by cold shoeing. This method enables farriers to go to the horse rather than the horse having to be taken to the smithy, and is thus more convenient. Hot shoeing, however, is said to be superior – the maxim being that the shoe is made to fit the foot, not vice versa, as in cold shoeing.

Whatever method is used, the farrier first removes the old shoe, cuts away the excess growth of horn, and rasps the surface of the foot to make it even. In hot shoeing, he then places the red-hot shoe against the bottom of the foot. It is hard to believe this is not a painful process, but, in fact, no pain is caused. The mark left by the hot metal tells the farrier whether he needs to alter the shape of the shoe, or needs to rasp the surface of the foot further. In cold shoeing, the farrier can only judge by placing the cold shoe against the foot. It is not as easy to alter a cold shoe by hammering as it is to alter a hot shoe, where the metal is more malleable.

In hot shoeing, the blacksmith cools the shoe, once satisfied with the fit. He then hammers it onto the foot using as few nails as possible, but enough to keep it in position; there are usually three on the inside and four on the outside. The points of the nails emerge through the front wall of the hoof and the farrier twists them off with the claw end of the hammer. Finally he hammers the nail heads against the hoof and rasps them smooth so they lie flush against the wall. The point at which the nails emerge in the front is critical; if too low, they may not keep the shoe in place, tearing down through the outer horn to cause cracking, while, if positioned too high, they can bruise the sensitive inner area of the hoof.

How often a horse needs new shoes will depend to some extent on how hard he is working and on what type of ground. However, most horses should be seen by the farrier once a month, since it may be necessary to cut back the excess growth of horn even if the horse is reshod with old shoes.

Even horses out at grass must be looked at by the farrier once a month. They will not move around sufficiently to keep the horn down to a reasonable length and, if it gets too long, it could cause them to trip. The horn, too, may begin to crack and break.

SHOEING – BEFORE AND AFTER

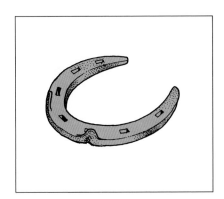

ABOVE: A horseshoe is held in place by nails driven into the tough, horny part of the foot. They are twisted off where they emerge higher up, and hammered down. A horse's shoes should never in any way interfere with the horse's natural actions and movements. Clips help to keep the shoe in place. These are small, triangular points that fit into the wall of the hoof. Usually there is one clip on the foreshoe and two on the hindshoe. Grip is extremely important and there are several methods for improving it, including calks and calking. However, studs are considered the most effective. These are not left in permanently, but screwed in place by the farrier.

It is absolutely vital that you watch for the signs indicating that your horse needs new shoes. Ideally this should be done on a regular basis, and the shoe and hoof should not be allowed to reach an unhealthy or dangerous state before you get new ones. BELOW: This horse is badly in need of new shoes. The indications are that the shoe itself has worn thin and rough and the clenches have risen and are standing out from the wall. All of these factors are dangerous and unhealthy for the horse.

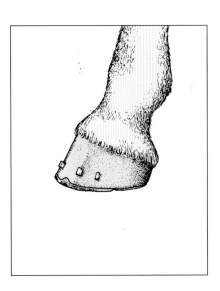

BELOW: A newly shod foot should show certain points indicating that it has been correctly fitted. The shoe should be fitted to the foot and not vice versa; the foot should be rasped and pared evenly and the frog in contact with the ground; a suitable number of nails should be used – never too many nor too few, and the clenches should be neat and evenly spaced. There should be no space between the shoe and the foot, and the clip should fit well on each shoe.

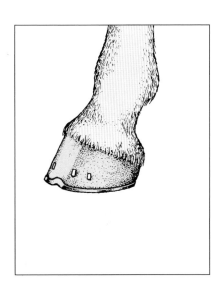

EXERCISE ROUTINES

All stabled horses must have regular and adequate exercise. Otherwise they can develop swollen legs, azoturia, and colic and will, in any case, be spirited and difficult to manage when ridden. They can also become bored and develop bad habits. The amounts needed vary with the type and weight of horse and the work it is expected to do; a hunter needs more exercise than a hack.

As with feeding, there are a few basic rules to remember. Most importantly, never exercise a horse until 1½ hours after a heavy feed; 1 hour after a small one. In any case, always remove the hay net an hour before exercise. Horses full of hay find breathing difficult when they are being worked hard.

The point of exercise is to get and keep the horse fit enough for the demands being made on it. A horse

SPECIMEN EXERCISE ROUTINE – based on a 16hh hunter

Exercise	Care and Management	Special Features
Week 1		
20 mins walking on the first day, increasing gradually to one hour	Gradually increase concentrated food, begin strapping	During the prework week the horse's feet must be checked and shod. All horses require one rest day each working week.
Week 2		
Walking for 1¼–1½ hrs over a 6–8 mile (9.5–13km) circuit	Check condition of legs and feet, watch for skin galls. Increase corn and vitamin supplements	Quiet lanes and roads with good surfaces are best for road work
Week 3		
Always walk the first ½ mile (0.8km). Then introduce very short periods of trotting, increasing their length gradually	Stable the horse at night and establish a regular routine	Schooling and longing in large circles can now be started
Week 4		
1½–2hrs work daily – split into schooling, longing, and road work. More frequent periods of trotting	Increase concentrated food	Trotting up gentle slopes can commence and increase slowly
Week 5		
After first walking and trotting, the horse may have a short, slow canter on soft ground. Then decrease pace gradually	Four feeds a day – increase concentrates and reduce time at grass. Reshoe if necessary	At this stage the coat should shine and the muscles should be hardening
Week 6		
A medium canter of reasonable length. Work at a sitting trot can now be started	Increase concentrated food ration. Maintain thorough strapping	Schooling can be intensified by trotting in smaller circles, and work at the canter
Week 7		
Canters can speed up. A short half-speed gallop may be added at the end of the week (on good ground). Jumping can begin	The horse will sweat and should wear a rug at night	The final phase of building up to full work. It is useful to introduce the horse to traveling and to company at this stage
Week 8		
On day 2 the horse can gallop at half speed up a gently slope. Always walk the final mile	Full rations of concentrate. The horse should gallop on alternate days, and do steady work on the others. Renew shoes	When the horse is fully conditioned thorough exercise must be maintained on days when it does not work

Exercise needs always differ, according to the size of horse, the type of work it is doing, and what it is being prepared for. Vary the routine accordingly.

brought up from grass, say, is likely to be in "grass condition." In such a case, fitness can be achieved only through a rigidly controlled program of exercise and feeding. Restrict exercise to walking, preferably on roads, for a week. Then combine walking with slow trotting. Soon, work can start in the school, while the period of road work can also be extended. Increase the amount of grain fed in proportion to the extra work. By the end of six weeks, the horse should be ready to be cantered over distances not exceeding ½ mile (0.8km). In the ninth week, it can have a gallop for up to ¾ mile (1.2km), but this should be strictly controlled so that the horse does not gallop flat-out at horse's-speed.

Indications of success are an increase in muscle and the disappearance of the profuse, lathery sweat of the out-of-condition horse. Never try to hurry the process; a horse cannot be conditioned through cantering and galloping, but only by slow, steady, regular work. This applies just as much to stabled horses and ponies.

Always aim to end the exercise with a walk, so that the horse comes back to its stable or field cool and relaxed. Once the tack has been removed, inspect the horse for cuts and bruises, pick out its feet, and brush the saddle and sweat marks. Then rug up or groom. If you have been caught in the rain, trot the horse home so that it is warm on arrival. Untack, and then give the

ABOVE: *It is important to introduce exercises to your horse gradually. Start with a slow walk, as here, then progress to the trot, and canter as and when you feel the horse is ready.*

horse a thorough rub down, either with straw or a towel. When this has been completed, cover the back with a layer of straw or use a sweat sheet. It is vital to keep the back warm to avoid the risk of colds and chills.

A thorough drying is essential if the horse is very hot and sweaty, but it will need to be sponged first with lukewarm water. Either restrict sponging to the sweaty areas — usually the neck, chest, and flanks — or sponge the entire body. Then, scrape off the surplus water with a sweat scraper, taking care to work with, and not against, the run of the coat. Next, rub down and, finally, cover with a sweat sheet. If possible, lead the horse around until it is completely dry.

Horses that have been worked exceptionally hard – in hunting, say, or in competitions – need further care. On returning to the stable, give the horse a drink of warm water. Then follow the procedures outlined above. Feed the horse with a bran mash and then leave it to rest. Return later to check that the animal is warm enough or has not broken out into a fresh sweat. Check for warmth by feeling the bases of the ears. If they are cold, warm by rubbing them with the hand and then put more blankets on the horse. If the latter, rub down and walk the horse around again until it is completely dry.

BELOW: *Longeing starts when a trainer feels that the horse is fit enough for a more extensive work program. The horse is led in circles around the trainer at the end of a lunge rein.*

LOADING AND TRAVELING

Careful planning when entering for a horse show, say, or going for a day's hunting, is essential if the horse is to arrive fit enough to undertake the tasks demanded of it. The first essential is to plan the journey; a fit horse can be hacked for up to 10 miles (16 km), walking and trotting at an average speed of no more than 6mph (7.5kph) (a grass-kept pony's average should not be more than 4 mph [5kph]). If the distance involved is greater than this, transport will be needed.

Horse boxes or car-towed trailers are the usual method of transport over long distances. Apart from the obvious mechanical checks that should be carried out before each journey, the horse's own requirements, too, need attention. A hay net is one essential; this should be filled with hay and given to the horse during the journey, unless the animal is expected to work hard immediately on arrival. Others include a first-aid kit; rugs (day and sweat); bandages; grooming kit; a head collar; a water bucket, and a filled water container. The last item is essential if the journey is to be a particularly long one, when the horse will need to be watered perhaps once or even twice *en route*.

In some cases – when hunting, for example – the horse can travel saddled up, with a rug placed over the saddle, but, in the case of competitions, a rug alone should be worn. Traveling bandages should always be used, as well as a tail bandage to stop the top of the tail from being rubbed. In addition, knee caps and hock boots should be worn as an added protection against injury while in transit.

Protection of the horse itself must start the night before, with an especially thorough grooming. Both mane and tail should be washed.

A grass-kept horse should be kept in for the night, if possible. The next morning, follow the normal stable routine, with the addition of a drawn-out strapping. Remember that, in the case of a show, the mane should be braided; this can be started the night before to ease the task of getting the mane into shape, but will need to be completed the following day.

Loading the horse

Getting a horse into a box or trailer is an easy enough task, provided that the process is tackled calmly and without undue haste. The simplest way is for one person to lead the horse forward, walking straightforward and resisting the temptation to pull at the head. A couple of helpers should stand behind the horse in case help is required, but out of kicking range.

The main reason for a horse showing reluctance to enter a box is usually its fear of the noise of its hooves on the ramp. This can be overcome by putting down some straw to deaden the sound. Loading another, calmer, horse first, or tempting a horse forward with a feed bucket containing a handful of oats, also act as encouragement.

A really obstinate horse, however, will have to be physically helped into the box. The way to do this is to attach two ropes to the ramp's rails, so that they can cross just above the horse's hocks, with two helpers in postion – one at each end of the ropes. As the horse approaches the ramp, they tighten the ropes to propel the animal into the box.

Unloading

To unbox a horse let down the ramp, untie the halter rope and, depending on the design of the box, push the horse's chest very gently to encourage him to step backward out of the box. In general, it is kinder to tether a horse outside the box (providing it is not in the glaring sunshine) or, better still, under the shade of some trees, than it is to leave him standing in the somewhat cramped conditions of most trailers. Only tether a horse, though, if you are sure he reacts well to such a practice and will not become excited if other horses at the show are ridden past.

ABOVE: *Lead your horse up the ramp; a layer of straw on the ramp will prevent his slipping. Avoid exciting your horse by shouting or waving your hands. Walk in with your horse, looking straightahead and tie him on a short rope. Close the ramp up quickly once inside. Provide your horse with a bowl of feed.*

HEALTH

Horses are tough creatures, but, like any animal, they can fall sick or be injured. A healthy pony or horse is alert, bright-eyed, and takes a keen interest in all that goes on around it. Ribs and hipbones should not be prominent, and the quarters should be well-rounded. The animal should stand square on all four legs. The base of the ears should be warm to the touch.

Signs of illness vary, but there are some general symptoms which can give warning of trouble to come. A field-kept pony which stays for a long time in one place, a horse which goes off its food, a willing horse which suddenly becomes "willful" – all these signs are indications that something is wrong. Other symptoms include: discharge from the eyes or nostrils; stumbling for no apparent reason; restlessness; dullness of eye or general lack of interest; sweating; kicking or biting at the flank; lameness; diarrhea; persistent rubbing of the neck or quarters against a wall or fence; apparent difficulty in breathing; coughing.

It is essential, therefore, to have a reliable veterinarian, and, if ever in any doubt, to call him without hesitation. Better to pay for a visit than to run the risk of mistaken self-diagnosis leading to a more serious illness, or even death. Nevertheless, all horse-owners should have a practical knowledge of first aid, and a first-aid kit is an essential part of any stable. It should be placed where it can be easily found in an emergency.

Nursing the horse

Like all animals, horses take time to recover from illness. The veterinarian will always instruct the owner in what to do, but, usually successful nursing is merely a matter of common sense.

AILMENT AND INJURY CHART
Skin & coat

Symptoms:	Causes:	Treatment:
Heat bumps (Humor) Various forms of size and shape. Rarely seen all over horse	Probably overheating from too much protein in system	Give bran mash with addition of two tablespoons of Epsom salts
Lice Itching, dull coat, appearance of small grey or black parasites on the coat	Unknown. Appears in spring on grass-fed horses or on animals which have been in poor condition and are now improving	Dust affected areas liberally with delousing powder, obtainable from veterinarian. Keep grooming kit separate
Ringworm Usually circular, bare patches on the skin of varying sizes which may or may not be itchy	Fungus infection which is highly contagious	Apply tincture of iodine to affected parts. Disinfect rugs and sterilize grooming kit. Keep horse isolated
Sweet itch Extreme itchiness of areas around mane and tail, apparent only in late spring, summer and early fall	Unknown, probably an allergy	Apply calamine lotion to relieve itching. Keep mane and tail clean. Lard and sulfur applied to the area can be soothing. Consult veterinarian
Warbles Maggot of the warble fly	Painful swelling on back	Bathe in warm water which will keep the lump soft and help to "draw" the warble from a small hole on the top of the swelling. The maggot can be gently squeezed out, but do not do this before consulting veterinarian

Digestive system
Colic Severe abdominal pain, characterized by pawing of the ground, restlessness, sweating, rolling, lying down and getting up, kicking, biting and looking at the stomach, groaning, cold ears	Poor or irregular feeding, wrong sort of food, exercise or drinking immediately after food, too much food when horse is tired. Worm infection	Call veterinarian immediately, meanwhile do what you can to relieve the pain. Keep horse warm, apply hot-water bottle to belly. Try to discourage horse from lying down or rolling
Diarrhea Very loose, watery droppings	Excessive fresh grass. Worms	Mix dry bran with food or add kaolin. Feed with hay. If persistent, call veterinarian
Worms Loss of condition, despite careful feeding	There are several types of intestinal parasites collectively known as worms	Regular doses of worming powder or paste, together with regular maintenance of pasture

AILMENT AND INJURY CHART
Feet

Symptoms:	Causes:	Treatment:
Bruised sole Lameness. Horse may ease the weight of the foot when at rest	Bruising by stones or rough-going and hard ground	Rest. If necessary, new shoes. Keep farrier informed
Corns Lameness. Heat in foot. Horse more lame on turn than straight	Ill-fitting shoe causing pinching. Shoe which has moved in. Bruising	Call blacksmith or veterinarian to cut out corn and advise further treatment
Laminitis (Founder) Obvious pain in the feet. Horse is reluctant to move and stands with its front feet pushed forward and its hind legs under it so that its weight is taken on the heels. May shift weight from one foot to another. Possibly a high temperature. Always apparent in front feet first but may affect all four feet	Overfeeding and not enough exercise. Grass-fed horses are especially prone to the disease after eating excessive amounts of new, spring grass. The feet become engorged with blood and the sensitive laminae in the hoof become inflamed and .y separate	Call veterinarian at once as prompt treatment can help the condition considerably. In the meantime cool the feet in running water from a hose and try to get the horse to walk, as exercise helps the feet to drain. Remove from grass and give light starvation diet
Nail blind Lameness soon after the horse has been shod	Shoe nail driven too close to the sensitive areas of the foot	Call blacksmith, who will remove nail and replace it correctly
Navicular disease Intermittent lameness, usually slight, followed by pointing, in which one forefoot is rested in front of the other on the toe. Gradual increase in tendency to stumble. Later, foot will contract at the heels	May be hereditary. Otherwise probably due to jarring of the foot through excessive road work or strain in hunting and jumping. This brings on lesions of the navicular bone	Consult veterinarian
Overreach Cuts and bruises to bulbs of heel	Toe of hind shoe hitting front heel	Bathe wound in salt solution. Call veterinarian if wound is severe. Prevention is better than cure – horse should be fitted with overreach boots
Pedal osteitis Intermittent lameness, later permanent	Severe jarring, brought on by too much road work or by jumping when the ground is very hard. This leads to inflammation of the pedal pone and bony growths on the bone	Rest. Bathing foot in cold water. Special shoeing may help
Quarter crack Crack or split in the wall of the hoof, extending upward into coronet	Mineral deficiency which makes hoof unusually brittle	Consult farrier who may fit special clips to hold edges of crack together, or put on special shoes
Quittor Lameness. Infection breaking out around coronary band	Infection in the hoof working its way upward to form abscess	Consult veterinarian
Seedy toe Revealed when trimming the hoof during shoeing. The outside of the hoof wall appears normal but a cavity is revealed when the horn is pared away	A legacy of laminitis. Tight shoes may also be a cause	Call farrier who will pare away the damaged horn. Then treat liberally with Stockholm tar

Head

Blocked tear duct Tears running down face	Sand, grit, or mucus causing blockage of tear duct	Call veterinarian, who will probably clear blockage by using a catheter to force sterile liquid through the duct
Broken wind Persistent cough, rapid exhaustion, double movement of flank	Breakdown of air vessels in the lung from overworking the horse	Incurable, may be alleviated by keeping horse out, work gently, and dampen food
Catarrh Thick, yellowish discharge from the nostrils	Inflammation of mucous membrane. May be cold infection preceding cough or allergy. Beware of infecting other horses	Clean nostrils with warm boric solution and smear with petroleum jelly. In summer, turn out to grass
Coughs and colds Thin discharge from nostrils; coughing	Infection; sometimes dusty hay or allergy	Isolate animal and keep warm; give regular doses of cough medicine. Consult veterinarian
Influenza Lethargy, cough, high temperature. Horse refuses food	Virus infection	Isolate. Keep warm. Rest. Call veterinarian. Prevention by inoculation is possible
Strangles Similar to those of influenza, plus swelling of lymph glands under the jaw, which eventually form abscesses	Contact with infected animal or with contaminated grooming kit, feed buckets, etc	Isolate. Call veterinarian. Feed hay and bran mashes and keep horse warm. Rest is essential

AILMENT AND INJURY CHART
Legs

Symptoms:	Causes:	Treatment:
Bog spavin Swelling in the front of the hock and on both sides at the back	Excess fluid in hock joint	None. Bog spavins look unsightly but cause no trouble
Brushing Sudden acute lameness. Injury around the fetlock joint	One leg striking against the other	Rest and hosing the affected part with cold water. Prevent recurrence by fitting brushing boots. Consult blacksmith
Capped elbow Swelling on the point of the elbow, level with chest. If infected, horse may be lame	Persistent irritation or rubbing of the elbow when lying down or because bed is too thin	Cold poultice the swelling and call veterinarian if infected. Special shoeing and provide thicker bed
Capped hock Similar swelling to capped elbow, but this time on hock. It is usually permanent but rarely painful	Knock or kick on affected area	None. Cold poulticing sometimes helps to reduce swelling
Cracked heels Sore patches, often suppurating, and deep cracks on the heels at the back of the pastern	An irritant in the soil which affects the heels and legs after they have been covered in mud. White legs are more prone to the condition	Apply ointment, using one with a cod-liver oil or zinc oxide base. Alternatively, dry poultice with dry warm bran, and bandaging
Curb Lameness. Outward bowing of line from point of hock to cannon bone	Sprain to ligament connecting point of hock with cannon bone	Rest. Cold poulticing application of liniment
Ringbone Lameness. Swelling of pastern	Blow, sprain or jarring which causes extra bone to form on first or second pastern bones, or both	Rest. Seek professional advice and be prepared for pony to be permanently lame
Speedy cutting Sudden fall or lameness. Cuts or bruises just below knee	One leg interfering with another	Rest. Bathe affected part with cold water
Splints Lameness. Heat and swelling in affected leg	Formation of bone between the splint and cannon bone	Rest. Cold water poultices. Once splint has formed, lameness disappears, leaving a permanent lump
Sprained joints and tendons Heat and swelling. Lameness in some cases	Jarring. Twisting of joint. Inflammation of the tendon	Cold water dousing. For tendons, pressure bandages. Rest
Sprung tendon Bowed tendon	Sprain to the tendon	None. This is an indication of a former sprain
Thoroughpin Swelling just above the hock which can usually be pushed from one side to the other	Strain	Pressure bandage to reduce swelling. Keep soft by massage or by applying goose grease

Giving medicine, for example, can present problems. The simplest way is in the feed, provided that the medicine is suitable and the horse is eating. Soluble medicines can be mixed in with the drinking water. Otherwise the veterinarian will advise.

The golden rules of nursing are gentleness, cleanliness, and the ability to ensure the horse's comfort and rest. When treating a wound always try to reduce the amount of dust in the stable. Decrease concentrated foods for a horse suddenly thrown out of work by lameness and substitute a mild laxative instead. Gently sponging eyes and nostrils will help refresh a horse running a temperature.

Care when old

Horses and ponies are frequently remarkably long-lived. Some ponies, for instance, are still leading useful lives at thirty, but caring for an elderly horse presents its own set of problems.

Teeth must be regularly filed (rasped), as the molars will probably become long and sharp if left untreated. Select the diet carefully; boiled barley, broad bran, chaff, and good quality hay form the best mixture for an old horse.

Eventually, though, some horses just lose interest in life. If the luster goes out of the eyes or the appetite wanes for no apparent reason, it is then kinder to have the horse put down. A veterinarian will arrange this. A humane killer is used, and death is instantaneous.

ABOVE: *Regular exercise is essential in keeping a horse fit and in such a condition that it can do the work expected of it without risk of injury.*

RIGHT: *This horse has sustained a surface wound on its leg. However minor the injury is, it is important to ensure it is thoroughly cleaned before dressing it.*

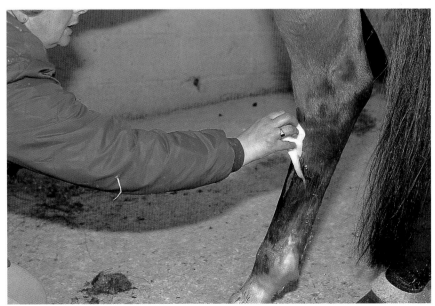

CHAPTER 2

TACK AND EQUIPMENT

SADDLES

The saddle evolved after the bridle. It came into being in order to make riding more comfortable for both rider and horse. Riding bareback is tiring and it is also somewhat insecure for the rider.

All designs of saddle, therefore, have one thing in common – the central gullet directs pressure off the spine and instead distributes the rider's weight evenly on either side of the back over the fleshy, muscular area covering the ribs. They also all assist a rider to control his or her mount by giving a firmer position from which to issue the aids.

The key feature of any saddle is a framework known as the tree, around which all saddles are built. The tree is usually made of lightweight wood. When this has been shaped and glued together, it forms the basis upon which webbing and padding will be bound and attached. Together, they determine the finished shape and, therefore, the purpose of the saddle.

The most widely used saddle in European riding today is the all-purpose saddle. This evolved from the riding style known as the Italian forward seat, which was developed some 60 to 70 years ago. The rider is positioned over the horse's center of gravity, shifting his or her position as the center shifts. An obvious example is leaning forward when a horse is jumping. Before this style was developed, a rider sat well back in the saddle, his legs pushed forward; if he had to jump, he would lean further back.

The design of saddles then in use encouraged such a position, so, to further their new ideas, the Italians developed a saddle that encouraged their new style. As this method of riding became widely adopted and practiced, most nations began to produce their own version of this saddle and now there is an enormous range of them.

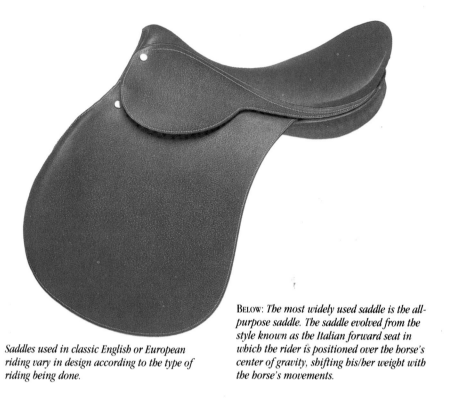

Saddles used in classic English or European riding vary in design according to the type of riding being done.

BELOW: *The most widely used saddle is the all-purpose saddle. The saddle evolved from the style known as the Italian forward seat in which the rider is positioned over the horse's center of gravity, shifting his/her weight with the horse's movements.*

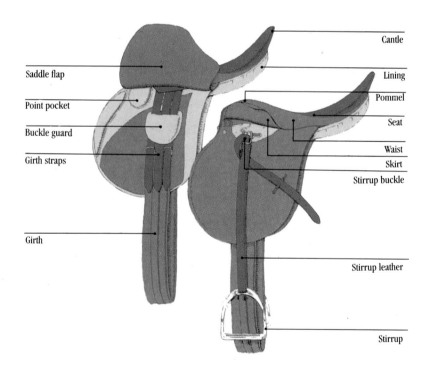

Saddle flap
Point pocket
Buckle guard
Girth straps
Girth

Cantle
Lining
Pommel
Seat
Waist
Skirt
Stirrup buckle
Stirrup leather
Stirrup

Sidesaddles

Sidesaddle riders are faced with a much smaller range of saddles – if, indeed, they can find one at all. Nearly all sidesaddles are secondhand, since, until the revival of the last decade or so, the art of aside riding – and thus the art of making sidesaddles – had almost died out. Most sidesaddles were made to fit a specific rider and this is an important point to bear in mind when buying a secondhand one.

PUTTING ON AN ALL-PURPOSE SADDLE

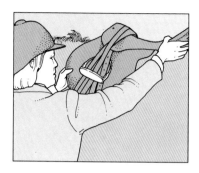

1 Place the saddle in position, so the pommel is just behind the withers, with irons up and girth across it.

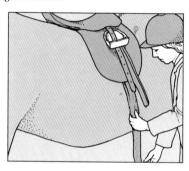

2 Check that the coat is smooth under the saddle flap. Then move around to the off side of the horse and let down the girth.

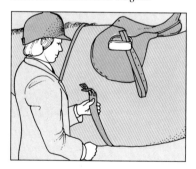

3 Return to the near side and pass the girth under the horse's belly. Check carefully to make sure that it does not get twisted.

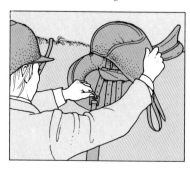

4 Buckle the girth to hold the saddle securely in position. Adjust stirrup leathers and tighten the girth again. Then mount.

PUTTING ON A SIDESADDLE

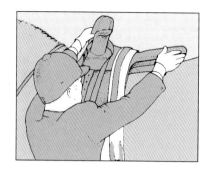

1 Standing by the horse's near-side shoulder, lift up the saddle and place it on the horse slightly forward of the withers.

2 As the girths, surcingle and balance strap are all attached to the near side of the saddle, undo these and take them down.

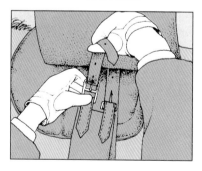

3 While still on the near side of the horse, check to make sure that the girth is buckled sufficiently high up on the saddle.

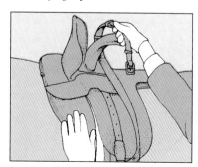

4 Unhook the surcingle from the pommel. If your horse is jumpy, ask an assistant to hold the saddle. It is not yet secured on the horse.

5 The girth, surcingle, and balance strap are now hanging down on the horse's near side and must be brought over to the off side.

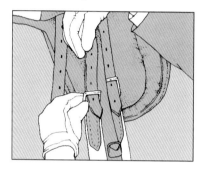

6 Bring the surcingle and girth through and buckle them. Make sure that the hook on the saddle flap is secured on the off side.

7 Pull the balance strap under the horse's belly, slotting it through the keeper on the girth. This will help to keep the strap in place.

8 Mount and ask an assistant to adjust the girth, balance strap, and surcingle, in that order, on the off side of the horse.

Western saddles

There are almost as many designs of Western saddle as there are European ones. The basic design features – the high horn, deep, wide seat and high cantle – were born out of practicality and are present to some extent in all Western saddles. However, over the generations, they have been subjected to endless variations and modifications.

The first stock saddle was said to have originated in Mexico, from whence it traveled to Texas, where it was copied. It continued to spread across the country, as ranchmen and cattle began to invade what had previously been buffalo territory. Alterations were made in different places to the rigging (the system of straps and girths that keeps the saddle in place), the horn and the swells (the padded area directly in front of the rider's knees which

varies in degree according to taste and the rider's job). In addition, tremendous variations occurred in the actual esthetic appearance of the saddles, according to the extent and elaborateness of the tooling on the leather and the addition of ornate silver buckles and other adornments.

The cowboy's working stock saddle on the home range differs from that used for Western pleasure riding, or more classic Western riding. Similarly, it differs from the saddle used by another ranchman, whose job involves him spending long hours of each day riding over the range. The cowboy working cows – roping

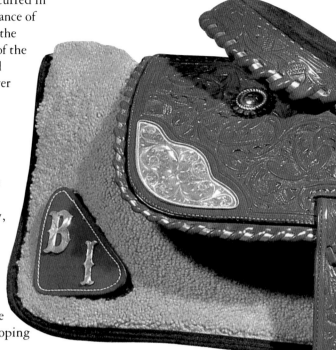

them and jumping out of the saddle hundreds of times a day – uses a saddle with a very strong, reinforced horn that can withstand the strain of a twisting, bucking cow. The saddle also has almost flat swells, so that these do not get in his way as he leaps quickly from his horse's back. The ranchman riding the range prefers a saddle with very pronounced swells and a deep seat, which makes the long hours in the saddle a great deal more comfortable.

LEFT: *The Western saddle was designed for practicality. While some are very elaborate with detailed leatherwork, most, like this one, are simple, sturdy, and functional. The horn is designed to withstand the stress of roping cattle and most saddles will have a deep seat and pronounced swells that allow the rider to sit relatively comfortably for long hours out on the range.*

Unlike European saddles, the Western, or stock, saddle, has no padding. This is because the extreme heat of the climate, which would often cause the horse to sweat profusely, would soon affect the padding by shifting it about and making it hard and lumpy. For this reason, Western saddles are always worn with pads and blankets underneath (these often were used as bedding for the rider when he had to sleep out on the range).

Once again, though, the saddle must still fit the horse correctly; an ill-fitting Western saddle can no more be made to fit a horse by putting extra blankets underneath than a European saddle can be.

LEFT: *Western saddles are much heavier than English or European saddles. Because they were designed to carry a person long distances, they were often padded for additional comfort. In putting on a Western saddle, begin by making sure that the hair around and under the saddle is lying flat and smooth. Standing on the near side, fold a blanket in half or thirds and put it high up on the withers, making sure it is evenly placed across the back. Place the saddle pad towards the front and position the saddle, pulling the blanket and pad up into the gullet so that they are not drawn too tightly across the horse's back.*

PUTTING ON A WESTERN SADDLE

1 Go to the other side and take the cinch and other straps down from the side of the saddle.

3 Then buckle or tie the cinch. The rear cinch does not have to be secured as tightly as the main cinch.

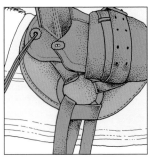

2 Return to the near side and pull these through under the horse's belly.

4 Finally, pick up each of the forelegs in turn and pull them forward to make sure the cinch is not pinching the skin behind the elbow.

WESTERN SADDLE VARIATIONS

BELOW: *This diagram shows the parts of the Western saddle.*

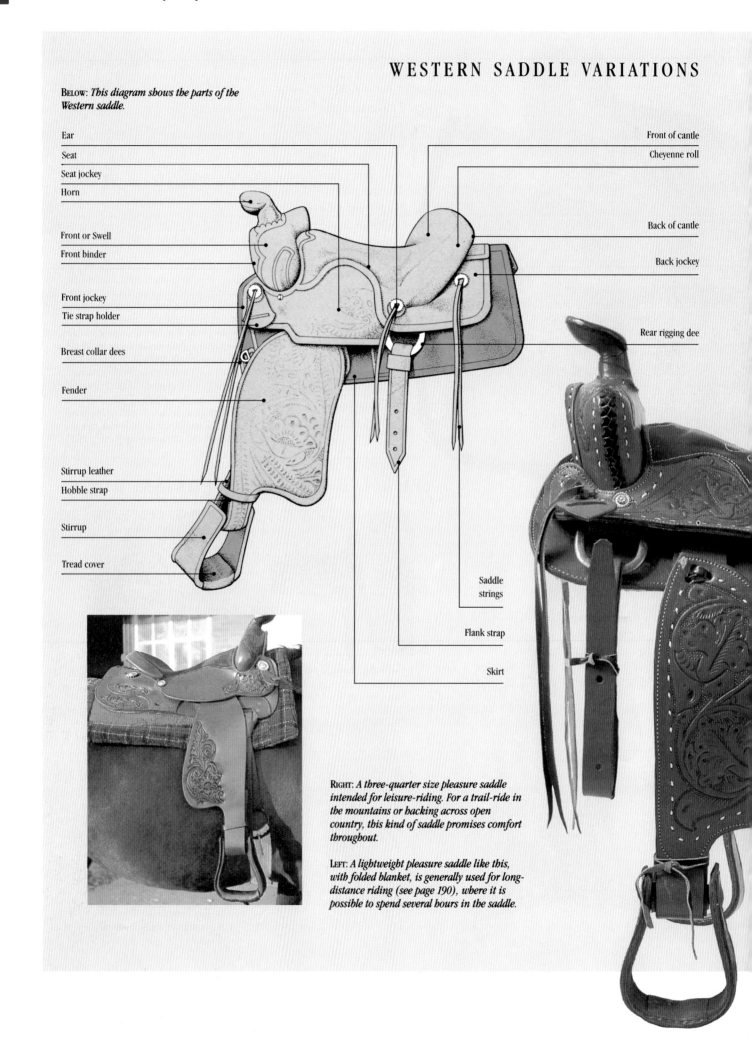

Ear

Seat

Seat jockey

Horn

Front or Swell

Front binder

Front jockey

Tie strap holder

Breast collar dees

Fender

Stirrup leather

Hobble strap

Stirrup

Tread cover

Front of cantle

Cheyenne roll

Back of cantle

Back jockey

Rear rigging dee

Saddle strings

Flank strap

Skirt

RIGHT: *A three-quarter size pleasure saddle intended for leisure-riding. For a trail-ride in the mountains or backing across open country, this kind of saddle promises comfort throughout.*

LEFT: *A lightweight pleasure saddle like this, with folded blanket, is generally used for long-distance riding (see page 190), where it is possible to spend several hours in the saddle.*

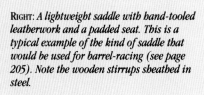

RIGHT: *A lightweight saddle with hand-tooled leatherwork and a padded seat. This is a typical example of the kind of saddle that would be used for barrel-racing (see page 205). Note the wooden stirrups sheathed in steel.*

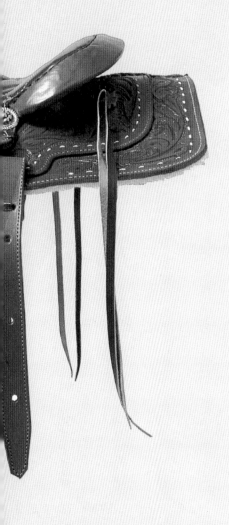

BITS

The snaffle is the simplest type of bit. Its mouthpiece may be jointed or unjointed, smooth or twisted, straight or curved, according to its individual design and purpose. At each end, it has a circular or D-shaped ring to which the reins are attached. Depending on how the rider uses his or her hands, the bit can be used to exert pressure on the outside of the bars of the mouth, the tongue and the corners of the mouth.

The mouthpiece of a curb bit can also be jointed or unjointed; alternatively, it may have a raised section in the center known as a "port." It may be "fixed," in that it is rigidly secured at either end to the cheekpieces, or movable which means the cheekpieces can move a little, independently of the mouthpiece. The slight flexibility this affords tends to lessen the severity of the bit. It also encourages the horse to play with it in his mouth, which helps to create saliva. This, in turn, helps to protect the sensitive tissues of the mouth.

The curb bit has long cheekpieces extending down either side of the mouthpiece, with the rings to take the reins placed at the bottom. In addition, a curb chain or strap, resting in the "curb groove" under the horse's chin, is fastened to either side of the bit. The lever action produced by the reins exerts pressure on the horse's poll through the headpiece and also on the curb groove through the curb chain or strap. The mouthpiece of the bit exerts pressure on the tongue, the bars and – in the case of very high ports – the roof of the mouth.

ABOVE RIGHT: *The bit contributes to the overall control of the horse and is used in conjunction with a number of other aids. The pressure exerted by the rider on the horse's mouth sends instructions and the horse's reaction should be one of relaxation and not of fear or pain. There are numerous types of bits ranging from the simple to the complex; however, it is still the skill of the rider which will determine the horse's movements and cooperation.*

TYPES OF BIT

ABOVE: *1 German snaffle; 2 straight bar snaffle; 3 rubber snaffle; 4 eggbutt snaffle; 5 Fulmer snaffle; 6 gag-bit; 7 loose-ring snaffle;*   *8 Kimblewick; 9 flexible rubber mouth Pelham; 10 Weymouth bridoon and curb bits; 11 Pelham bit; 12 Scamperdale*

HOW THE BRIDLE AND BIT WORKS

To understand how and why the bridle enables the rider to control and direct his horse, it will help to look at the parts of the animal's head that are affected. The bit and bridle can be used to exert pressure on eight parts of the head.

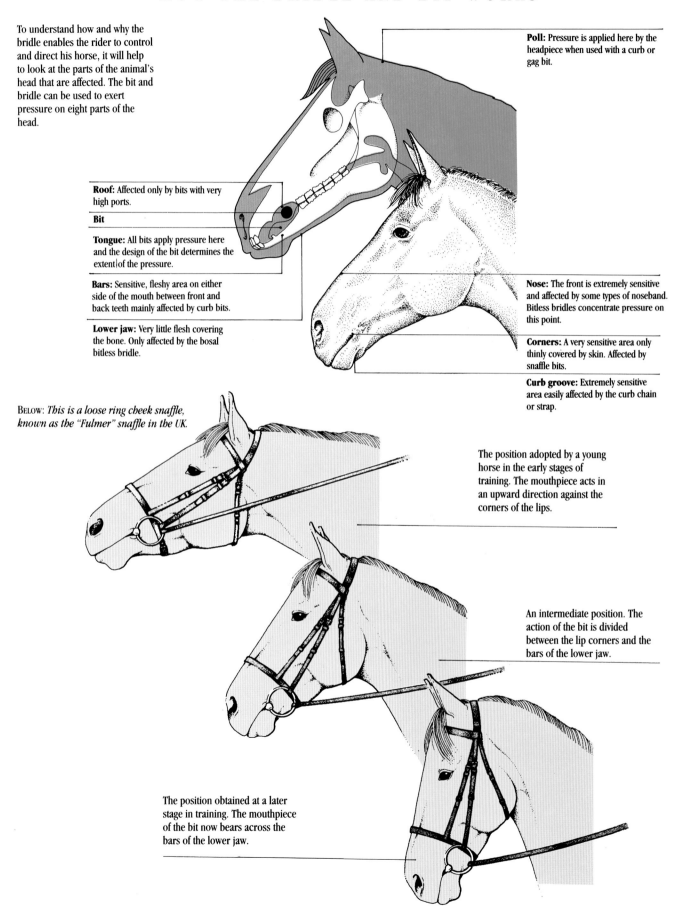

Poll: Pressure is applied here by the headpiece when used with a curb or gag bit.

Roof: Affected only by bits with very high ports.

Bit

Tongue: All bits apply pressure here and the design of the bit determines the extent|of the pressure.

Bars: Sensitive, fleshy area on either side of the mouth between front and back teeth mainly affected by curb bits.

Lower jaw: Very little flesh covering the bone. Only affected by the bosal bitless bridle.

Nose: The front is extremely sensitive and affected by some types of noseband. Bitless bridles concentrate pressure on this point.

Corners: A very sensitive area only thinly covered by skin. Affected by snaffle bits.

Curb groove: Extremely sensitive area easily affected by the curb chain or strap.

BELOW: *This is a loose ring cheek snaffle, known as the "Fulmer" snaffle in the UK.*

The position adopted by a young horse in the early stages of training. The mouthpiece acts in an upward direction against the corners of the lips.

An intermediate position. The action of the bit is divided between the lip corners and the bars of the lower jaw.

The position obtained at a later stage in training. The mouthpiece of the bit now bears across the bars of the lower jaw.

NOSEBANDS

In almost every instance the noseband fulfills some practical function connected with increased control of the horse. The only exception is the plain cavesson noseband which, unless it is used as an anchorage for the standing martingale, serves no more than an esthetic purpose. In its normal position the cavesson is fitted so that two fingers can be inserted between it and the jawbone. If fitted tightly and a little lower than usual it can partially close the mouth, but not to the same extent as the drop noseband.

The drop noseband

The drop noseband is without doubt the most important type of noseband in the context of modern riding, whether it is being used in the schooling of a horse or in competition. The nosepiece is fitted some 3 in (7.5 cm) above the nostrils, just below the termination of the facial bones. The back strap is then secured under the bit, so as to lie in the curb groove. Positioned in this way and adjusted fairly tightly, the noseband stops the mouth from being opened and, as a result, also prevents the horse from evading the action of the bit in that manner. For the same reason, its use ensures that the bit remains central in the mouth, since the horse cannot slide the bit over to one side or the other.

The pressure of the noseband therefore

ABOVE: *Drop noseband, fastening below the bit, a very common modern bitting arrangement*

assists and strengthens the action of the bit. Pressure on the rein is transmitted to the nose as the horse's lower jaw gives to the bit action. In turn such pressure causes the horse to drop his head, allowing the bit to bear across the bars of the lower jaw, in which position it will have the greatest effect. However, it is also the case that the pressure exerted on the nose — if sufficiently strong — can cause a momentary check to the breathing, which will contribute to the dropping of the head.

Variations on the basic drop noseband also exist. One widely-used type is called a "Flash" noseband, after a jumper who wore it, and another is the Grackle, Figure 8, or "cross-over" noseband. The former is designed for use with a standing martingale, the two crossing straps sewn to the center of the cavesson, which fastens under the bit, being the means by which the mouth is kept closed. It is not, however, as effective in lowering the head as the straightforward drop, since the point at which pressure is put on the nose is higher than in the true drop noseband. Far less nose pressure can be applied as well.

The Grackle was named after a horse of that name who wore one when he won the British Grand National in 1931. In fact, the Grackle has lost the chief features of its design in recent years and has become merged into the general concept of a cross-over noseband. In its general form, the top strap, fastening above the bit, was carefully shaped, so that it and the lower strap, fastening under the bit, were kept exactly in place; this being assisted by the connecting strap at the rear. Nose pressure was localized at the point of intersection of the straps, but could be adjusted at the headpiece so the point was raised.

The cross-over nosebands are possibly not as precise as the conventional drops, but they are probably more suited to some horses. This is because of their reduced degree of restriction,

DIFFERENT TYPES OF NOSEBAND

Australian racing cheeker

Grackle or Figure 8 noseband

Sheepskin noseband derived
from the harness racing shadow noseband

Kineton or Puckle noseband

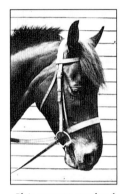
Plain cavesson noseband

Flash noseband

A raised, show-type noseband
with a snaffle bridle

particularly in relation to the respiration, which make them suitable for the hard-pulling cross-country horse and for the steeplechaser.

The Kineton, or Puckle, noseband goes to the opposite extreme. It has no pretensions to be other than a strong stopping agent for use on very hard-pulling horses. It makes no attempt to close the mouth, the metal loops, fixed behind the bit, transmitting the very considerable pressure which can be obtained directly to the nose by means of the nosepiece which is adjusted both low and fairly tight. The nosepiece is frequently reinforced with a core of light metal.

A noseband popular in racing and other equestrian circles is the sheepskin-covered noseband, which originally was used with harness trotters as an "anti-shadow" or "anti-shy" noseband. In the context of the trotting horse the "shadow roll," as this noseband is termed, has a

definite purpose. In conjunction with the characteristic extended nose position of the trotter moving at speed, it prevented the horse from seeing shadows on the track or variations in the surface color which might cause him to check or break his gait. It has far less to commend it in the context of the riding horse, and very little in the context of racing, either. No firm opinion, for example, is held as to whether the use of the sheepskin-covered noseband is supposed to encourage a horse to put his head up or down.

The Australian Cheeker

A final, useful piece of equipment is the noseband referred to as an Australian Cheeker, which, for no very good reason, is confined largely to the racing scene. Usually made of

rubber, the cheeker fits over the bit rings on either side and then joins into a central strap which runs right up the face and fastens to an attachment on the headpiece of the bridle, right between the horse's ears.

Correctly fitted, the noseband lifts the bit in the mouth, which makes it more difficult for a horse to get his "tongue over the bit", thus evading its action and causing even more serious trouble by "swallowing" the tongue. This, however, is not its only effect. For some reason which is still not satisfactorily understood, anything running up the center of a horse's face exerts some form of psychological restraint and is a very effective ploy to use in the case of hard pullers. In more elaborate forms the system can be seen incorporated into the Rockwell bridle and the Norton Perfection, or Citation; these are bridles of American origin used in racing.

BRIDLES

The various parts of the bridle are virtually common to all types, although they may not look the same or employ the same method of fastening.

The head of the bridle, passing over the horse's poll, has, attached to it, the cheeks to which the bit is secured. The throatlatch (pronounced throatlash) is usually incorporated in the head, though, in some instances, it is a completely separate strap attached to the head by a loop fixed between the horse's ears, as in some American patterns.

Certain types of bridle, however, omit the throatlatch completely. There are, for instance, no throat-latches on the bridles used in the Spanish Riding School in Vienna. The reason for this is that a throat-latch, if it is too tight, can discourage a horse from flexing at the poll, because of the discomfort it would cause. The Spanish School Lipizzaners are, in any case, naturally thick through the jowl and, since they are unlikely to get into situations in which the bridle may be pulled off, there is no need for a throatlatch. Its other purpose is to prevent the bridle from coming off in the event of a fall, for instance. A number of Western bridles also dispense with the throatlatch, preferring to keep the bridle in place by a slit passed over either both ears or a single ear.

The browband, or "front" as it is sometimes known, is fastened by loops to the headpiece and serves to keep the latter from sliding backward. There is then, in most bridles, but not in all, a noseband and then finally a pair of reins.

RIGHT: *Parts of a double bridle. The extra pair of reins is also a requirement of the Pelham and Gag bridles.*

ABOVE: *The double bridle, comprising curb bit and bridoon (i.e. a light snaffle bit) generally goes under the name Weymouth.*

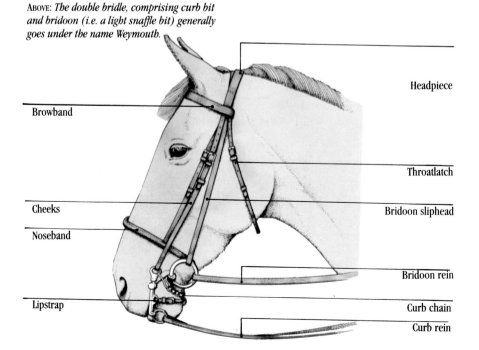

Browband

Cheeks

Noseband

Lipstrap

Headpiece

Throatlatch

Bridoon sliphead

Bridoon rein

Curb chain

Curb rein

Double bridles and Pelhams

The above describes the composition of a snaffle bridle, but, in the case of double bridles and Pelhams, additions are needed. On a double bridle, for instance, there has to be a sliphead, from which the bridoon is suspended. A sliphead is a strap and one cheekpiece passed through the loops of the browband under the headpiece. The cheek of the sliphead is placed on the off side so that its buckle matches that of the noseband on the near side.

In both the double and Pelham bridles, a pair of extra reins is necessary, the curb rein always being the narrower of the two. There is also the addition of a lipstrap, which is attached to the dees halfway down the cheeks of the bit and through the "fly" (flying) link in the center of the curb chain. Its purpose is to keep the curb chain in place.

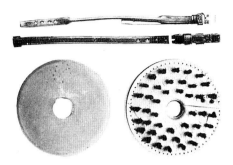

ABOVE: *(top) Two types of lipstrap, rounded and flat leather; (left) Rubber cheek guard; (right) Brush pricker used on one side of the bridle to prevent a horse from hanging in that direction.*

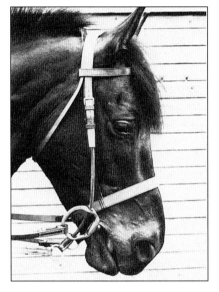

ABOVE: *Gag bridle*

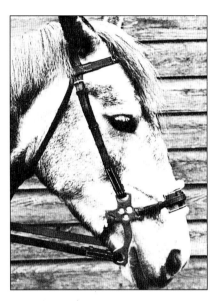

ABOVE: *Hackamore or bitless bridle*

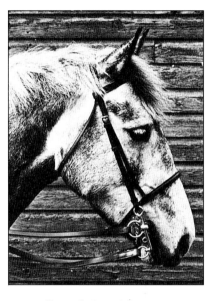

ABOVE: *Pelham, which can be converted to a single rein with a leather rounding joining the bridoon and curb rings.*

ABOVE: *Hunting snaffle*

ABOVE: *Fine leather, lightweight show bridle with raised and swelled noseband.*

ABOVE: *Weymouth bridle in hunting, general-purpose weight. Bridles can be attached to bits by sewing, hook studs or, very occasionally, buckles.*

MARTINGALES

All martingales share the common objective of assisting the action of the bit by restraining in one way or another the positioning of the horse's head and neck. In some cases they may go further by accentuating a particular action, or even by altering the character of the bit.

These auxiliaries can be regarded conveniently in two sections. There are those that seek in the simplest ways to prevent the head from being raised out of the control of the hand, and those which are intended as serious, long-term schooling aids, designed to affect the whole outline of the young horse or to attempt the correction of that of the older animal. In effect, the martingale is an aid to improved balance – although it would not be considered so by the purist.

Simple martingales

The simplest form of martingale is the standing or "fast" martingale. Essentially this is an adjustable strap attached at one end to the girth between the forelegs and at the other to the rear of a plain, stout cavesson noseband. To keep the strap in place it is fitted with a neckstrap, which can also be used by the rider in case of emergency to stay in the saddle.

The martingale prevents the horse, very effectively, from throwing up his head, and, depending on how it is adjusted, places the bit below the hand and acting squarely upon the bars of the lower jaw. Adjusted to a sensible length there is no reason why it should restrict the horse when jumping, since the head and neck are then stretched forwards and downwards – not upwards.

The martingale, in a very much strengthened form, is a virtual essential in the equipment of the polo pony, but sometimes appears as the more colorful *pugaree* martingale. This piece of equipment comes from India. It is, in fact, a length of colored turban cloth.

Occasionally, in some areas of Western riding, the standing martingale acts directly on the mouth, rather than the nose, being divided at the top end of the strap and the bifurcations being fitted with snap hooks to fasten to the bit. A similar device was once in vogue in Europe.

Running martingales

The running martingale is slightly more complex. Again, this has the central strap divided, the ends of the two branches being fitted with rings through which the reins are passed. When the horse attempts to throw up his head, the action is countered by pressure on the bars of the lower jaw through the bit. The tighter the martingale is adjusted, the greater is the restriction on the position of the head. In theory, it is recommended that the martingale is fitted so that the rings are in line with the withers, but in practical terms the adjustment is usually somewhat tighter. In show jumping, for instance, the rein frequently forms a distinct angle between the mouth and the hand. In both instances the martingales, whether standing or running, alter the action of the snaffle bit, since it must be presumed that without the restraint imposed by them, the horse's head would be held higher, so that the bit would act more upon the corners of the lips than otherwise. Using the martingales, the bit is placed over the bars of the lower jaw, the action

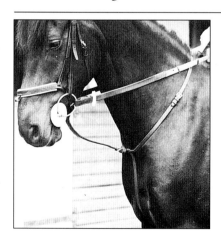

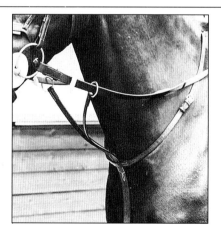

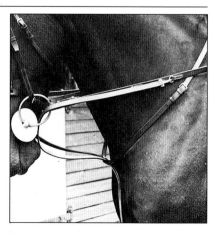

LEFT TO RIGHT: *Standing, or "fast" martingale controlling position of head by pressure on the nose; Running martingale, imposing control by pressure on the mouth. The "stops" in* *advance of the rings prevent the latter sliding forward and becoming caught on the bridle or over a tooth; The Market Harborough or German rein (in Germany often called the* *English rein), which is a much improved version of the old draw rein, the action being brought about when the horse evades by throwing his head above an acceptable level.*

being more direct in the case of the running martingale.

Racehorses very often wear a running martingale in which the branches are joined by a triangular-shaped piece of leather. This is called a "bib" martingale and is fitted as a precaution against an excited horse getting caught up or even getting his nose between the two branches. Racing trainers also use a small piece of equipment, which, despite being called the Irish martingale, is an intruder within the martingale family, since it has none of the group's familiar objects or characteristics. It consists of a short strip of leather joining two rings together, through which the reins are passed, and its use has no influence on the head position at all. Its purpose is to assist the correct direction of the rein pull and to prevent the reins being brought right over the horse's head in the event of a heavy fall.

It is usual to fit "stops" on the rein with which a running martingale is used. These are shaped pieces of rubber or leather which are slid on to the rein and fit tightly. Ideally they are positioned about 8–10in (20–25cm) from the bit. They prevent the rings of the martingale from running too far forward and becoming caught on the bridle in some way, or even caught over one of the horse's teeth.

In theory again, the running martingale should not be used with a double bridle, since the latter, on a schooled horse, provides, without outside assistance, all that is necessary to obtain the required head position and degree of control. In practice, however, the martingale is on occasion used with the bridle. In this instance, it should logically be placed upon the curb rein to assist the lowering of the head which is the purpose of the latter. As often as not, it is to be seen on the bridoon rein, when it becomes a

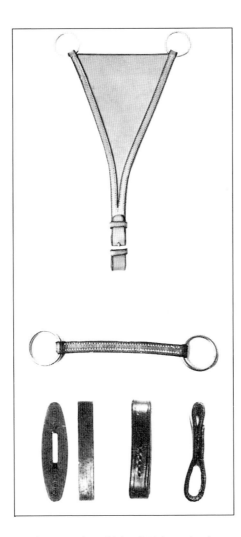

Bib martingale and (above) Irish martingale, and rein stops made from rubber and leather.

contradiction in terms; sometimes with both reins passed through the rings. This last may constitute an effective braking system, but makes a nonsense of the reasons why the double bridle is used by riders.

The German rein

Anothern pattern of martingale much used in showjumping goes under the name of Market Harborough or German rein in Britain, though, in Germany, it is often called an English rein. The rein derives from the more straight-forward draw rein which fastens to the girth on either side, then passes

through the bit rings back to the rider's hand. Though it is termed a rein, it is much more like a martingale in its action.

The martingale has two strips of leather attached to a ring on the chest and passing through the bit ring to a fastening on the rein. The adjustment is made either with a buckle or by clipping the leather strip to one of three or four small metal Ds set on the rein. The action is quite simple. So long as the horse carries his head acceptably, the rein operates in the normal fashion. A downward pressure on the mouth only occurs when the horse throws up the head, causing the leather strips to tighten and to pull down on the bit rings.

For the moment, outside of polo, the standing martingale is out of favor, although it is used as a training aid for jumping. The running martingale, however, is widely used in all competitive events, with the natural exception of dressage.

The pulley martingale

The older generations of horsemen and women seem to have given much more thought to bitting than is usual in these days of universal horsemanship based on the snaffle and drop noseband. One older type of running martingale such riders used was the pulley type, in which the rings were attached to a cord which passed through a pulley at the top of the body strap. Its advantage was that, in making sharp turns and so on, the horse was allowed to bend his head in the direction of the movement, without the kind of restriction on the opposite side of his mouth which is inevitable to some degree with the conventional pattern.

THE WESTERN HACKAMORE

In Europe, "hackamore" often is used mistakenly to describe a bitless-type bridle, usually fitted with metal cheekpieces. The true hackamore largely consists of a heavy, braided rawhide noseband, the shape of a *réal* tennis racket, with a large knot at the end which lies under the horse's chin. The noseband itself is called a bosal and is fitted to the horse by means of a lightweight latigo headstall. This may be slit at an appropriate point so that it can be kept in place by passing it over an ear, or may be made more secure by the addition of a browband, or *cavesada*.

The hackamore is completed by the addition of a rope made from mane hair, which is called the *mecate* and usually by a *fiador*, made from the same material or sometimes from cotton. The *mecate* is attached to the heel knot by a system of "wraps" to produce a delicately balanced and sophisticated control device, the heavy rope reins and the heel knot combining to act as a counterweight to the substantial

ABOVE: *A variation of the bitless bridle, misnamed hackamore, now in general use. It achieves its object by putting pressure on the nose. The sheepskin padding on the nose and rear strap is to prevent chafing. It is also necessary to vary the fitting frequently to avoid callousing the nose.*

nosepiece. The *fiador* is used as a throatlatch and adjusted short enough to prevent the heel knot from bumping annoyingly against the lower jaw as the horse moves.

From hackamore to bit

Initially, the hackamore is used with both hands, but, as the horse's schooling progresses, the reins are used in one hand only. The fully schooled hackamore horse can carry out all the movements required of him in a state of constant balance and at high speed. He can make the sudden stops, the pivots (the equivalent of the dressage pirouette, though not the same movement), the turns and the rein-backs all on a looping rein and without his mouth ever being touched. The final stage is the graduation from the hackamore to the bit, usually, but not always, a fairly long-cheeked, high-ported curb (the port is the inverted U in the mouthpiece which allows room for the tongue and permits the bearing surface of the bit to rest directly on the bars). This transition is a gradual one, made with the help of a much lighter hackamore fitted with a pair of very light rein ropes. It is often known as a two-rein bosal. In the final stages, control passes to the bit, the latter being supported by a bosal of the very lightest proportions acting independently without reins.

The finished Western horse is ridden in a light curb bit bridle without a bosal or noseband of any sort, and a floating, or looping, rein, which exerts no more than a minimal contact on the mouth. Sometimes the reins are weighted by the addition of small decorative pieces of metal, but the ideal is for the horse to ride on the weight of a plain ¼in (6mm) rawhide rein.

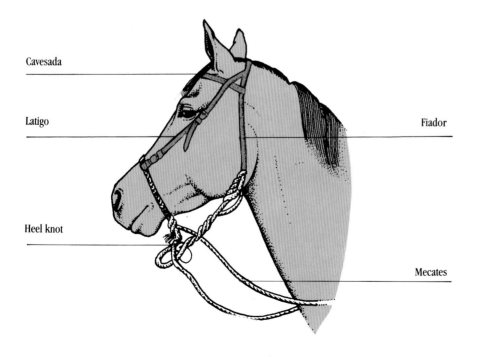

Cavesada

Latigo

Heel knot

Fiador

Mecates

LEFT: *The parts of the hackamore including the* fiador, *the throatlatch which prevents the heel knot from bumping annoyingly against the lower jaw.*

The European hackamore

The European equivalent of the hackamore is the variety of bitless-type bridles, deriving from the hackamore system. Of these, the best known is Blair's pattern. This bridle consists of the usual type of headpiece, a noseband, a curb, or back strap, and a pair of long metal cheeks to which the last two items are attached. Control is effected by exerting pressure on the nose and on the back strap embracing the lower jaw, the potential severity of the action being dependent upon the length of the cheek. Since nosepiece and back strap must be adjusted tightly to be effective, both must be soft and well-padded. The nosepiece should rest, as in the case of the bosal, above the ending of the nose cartilage, so as not to restrict the breathing, and its position needs to be altered continually.

Contrary to the general view, the bitless bridle is not suitable for novice use, since a novice could do far more damage with it than with a metal bit. Nor will it produce sudden and miraculous results. It is the precision tool of the expert horseman with a pair of delicate hands. Ideally, it, too, should operate from a floating rein, changes of direction being made by carrying the required rein outward and combining that action with a shift of body weight in the same direction. Less severe and often effective on a horse whose mouth, for whatever reason, precludes the use of a bit, are the far shorter cheeked bitless bridles, but they have little in common with the hackamore system.

An interesting bitless bridle is that perfected by William Stone, a lorimer in Walsall, Britain, who worked for many years with the firm of Matthew Harvey Ltd. It is called the W S Bitless Pelham and it, or something very similar, is still available today. The bridle employs two reins – hence the term "Pelham" – the top rein acting on the nose and the lower one on the curb groove by means of a curb chain. The metal cheeks of the bridle, which are comparatively short, move independently and thus allow a certain finesse in the action which is not found in other patterns in current use.

The advantages of the hackamore system are obvious enough in the schooling of polo ponies, for instance, but perhaps less so in regard to the modern, competitive, horse world. This is considered to be unfortunate by many, because there is much to commend to the present-day rider in this older and infinitely skillful school of riding.

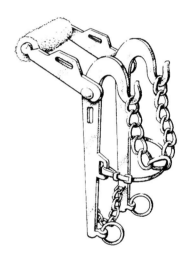

ABOVE: *The William Stone Bitless Pelham. It is relatively sophisticated and is used with two reins.*

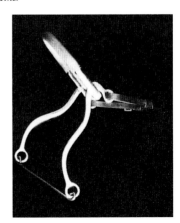

ABOVE: *Another bitless bridle, acting as a form of curb on the nose and employing a single rein.*

BELOW: *This diagram shows the action of the European hackamore.*

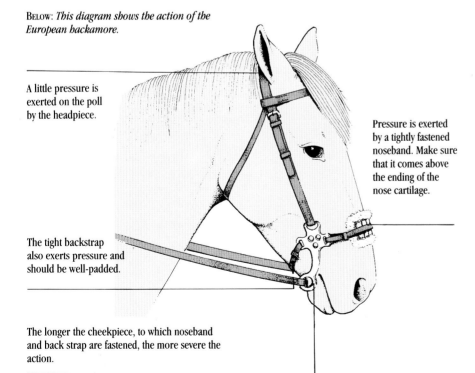

A little pressure is exerted on the poll by the headpiece.

Pressure is exerted by a tightly fastened noseband. Make sure that it comes above the ending of the nose cartilage.

The tight backstrap also exerts pressure and should be well-padded.

The longer the cheekpiece, to which noseband and back strap are fastened, the more severe the action.

RUGS

If a horse is worked during the winter, he will normally need clipping to avoid excessive sweating. The corollary of this is that his natural coat must be replaced by an artificial one, if he is not to catch cold. Winter stable rugs exist for this purpose and come in a number of varieties. The traditional pattern, made of jute or finely woven canvas and fully lined with a gray wool-mixture blanket, is probably the most successful. The rug is kept in place by a leather buckle fastening at the chest and either by a jute surcingle attached to the rug, a plain leather roller, or an anti-cast roller. The last is an adaptation of the ordinary leather roller, as it is fitted with a metal arch which goes over the horse's spine. It prevents the horse rolling over in his box and perhaps getting cast – that is, unable to get up unaided.

For added warmth, a "pure new wool" blanket can be placed underneath the rug. These blankets are traditionally fawn in color, with black, red, and blue stripes at either end. They weigh approximately 8lb (3.6kg). Lighter-weight blankets, made of "all wool" as opposed to "pure new wool" in similar colors, are also on the market, plus grey or brown wool-and-fiber mixture varieties.

An anti-sweat sheet, similar to a string vest, can also be used under the night rug for extra warmth. The sheet, made of cotton mesh, creates air pockets next to the body to insulate the horse against extremes of heat or cold. Its normal use is with a sweaty horse. For best results, it should be put on the hot animal next to the skin, with a day rug or summer sheet on top of it, to prevent the horse from getting chilled.

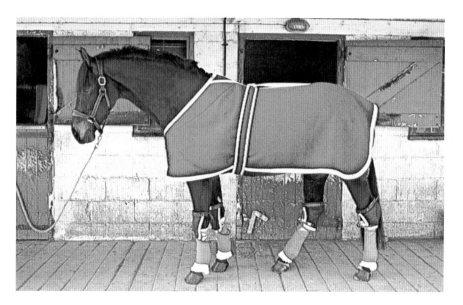

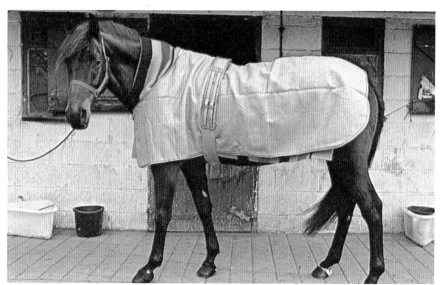

TOP: *A horse ready for traveling, wearing a headstall, wool day rug, roller, tail guard, kneecaps, hock caps and protective bandages.*

ABOVE: *A jute night rug worn over a striped underblanket and secured by a body roller.*

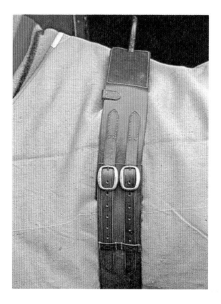

RIGHT: *Leather stable roller fitted with anti-cast hoop.*

Other rug varieties

A range of nylon quilted rugs, similar to the human anorak, are available. Most are made of nylon with polyester filling, with either a brushed nylon or cotton lining. They are extremely light and warm at the same time. These rugs are again kept in place by a surcingle or roller, with a nylon and metal fastening at the chest.

An alternative method of fastening is particularly popular in the USA. Here, the rug is fastened at the front with a chrome box clasp, with a cross-surcingle, designed on stress engineering principles, to keep the rug in place. The two webbing straps sewn on the rug equalize the tension by starting from the points of each shoulder, crossing under the horse in the normal roller position to the top of the hindquarters.

A rug, made of a very light fabric called Thermatextron, has a similar cross-surcingle fastening. Laboratory tests have demonstrated that this fabric has a higher degree of thermal insulation than any other. It also absorbs less moisture.

Both these factors mean that the heat generated by the horse's body is conserved. For the maximum benefit, the rug should be worn next to the skin. It can be put over a wet or sweating horse, the damp evaporating through the fabric.

A further type of rug with cross-surcingle fastenings is known as a "banner blanket." It is made of triple-thickness woven acrylic fabric and keeps horses warm and the dampness out, even in the cold US winters for which it was designed.

TOP RIGHT: *Mesh anti-sweat rug.*

CENTER: *Linen/cotton summer sheet with fillet string and light surcingle.*

BOTTOM: *Striped blanketing, originally used by the Hudson Bay Company for trading with the North American Indians, is traditional to the horse clothing industry.*

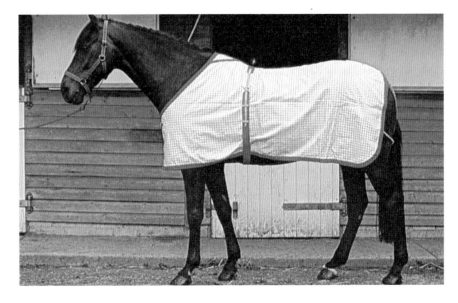

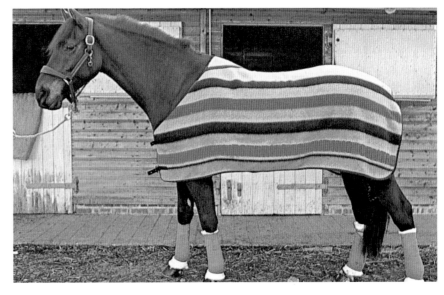

BANDAGES

Horses wear bandages for a number of reasons. Principally, bandages are used to protect the horse's tail and legs when traveling; to prevent injury to the limbs, should the horse knock himself in the field or when being ridden; as a support for the tendons during exercise; to keep the horse warm in the stable; and to hold dressings in place when veterinary treatment is required.

The type of tail bandage commonly used is made of a strip of elastic gauze, about 8ft (2.5m) long and 3in (7.5cm) wide. This is bound around the tail to keep the hair flat and in place. Such a bandage should always be worn during traveling to prevent the horse from breaking his tail hairs, if he rubs his tail on the back of a trailer or against the side of a box. The tail hair should be dampened before the bandage is applied; the bandage itself, however, should not be wetted, as it might otherwise shrink when in place, causing both discomfort to the horse and damage to the tail. It should be wound around the tail, from the top down to the end of the dock, and then up again for a few inches, before being secured.

The same 3in- (7.5cm-)wide gauze bandages are also used as exercise bandages to support the tendons. Here, they are usually worn over a layer of Gamgee tissue, thick cotton batting, pads of cotton, felt, or other similar materials. Such bandages are applied at the top of the cannon bone under the knee and extend down to above the fetlock joint. The tapes of the bandage are fastened securely on the outside of the cannon bone with a knot. Bandages fastened on the inside of the leg are more likely to come undone while the horse is working, since the knot can be caught by a blow from the opposite leg.

The purpose of the bandages is to help to absorb concussion and to make sure that the pressure ridges, which could damage the tendons, are evened out. When used on horses taking part in strenuous activity, such as competing cross-country, show jumping, or hunting, it is advisable either to sew the bandages in place after tying the tapes, or bind them round with surgical tape. This ensures that the bandage will not come undone, while, if the surgical tape is used, this will also supply a waterproof covering.

Support must be given to each pair of legs, if it is to be effective – that is, both forelegs have to be bandaged, or both hindlegs. The horse can be bandaged all around if this is considered necessary.

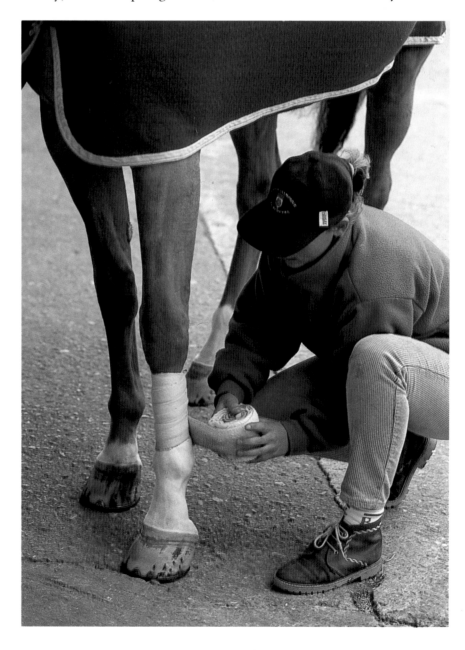

RIGHT: *This show jumper is bandaged in front for protection and support; the bandages are firmly secured by a wrapping of surgical tape. The tape covering also protects the bandages from the consequences of being immersed in water, which can cause the bandage to shrink and become uncomfortably tight. It is for this reason, as well as for that of security, that surgical tape is so frequently used in the sport of eventing.*

Gauze bandages can also be used to keep dressings in place, or as a cold water bandage for sprains and swellings. A type of proprietary elastic sock can also be used for this purpose. Another recent innovation is the "self-stick" variety of gauze bandage. Its use eliminates the need to use tapes.

Woolen stable bandages, approximately 5in (12.5cm) wide and 8ft (2.5m) long, are particularly useful for keeping a stabled horse warm in winter and protecting the legs of horses in a box or trailer. In the second case, it is particularly important to make sure that the bandage and its padding come down completely over the coronet, thus giving protection to the heel. Gamgee or similar padding is again used, though, in the case of some bandages, this is unnecessary. This type is made of thick, padded wool, with stockinette at each end.

The bandages are applied between knee or hock and continue down over the fetlock joint, this giving more warmth and protection than exercise bandages. They are fastened with strips of Velcro (as in the case of the padded bandages described above), or tapes. Velcro strips are quick and easy to apply, but their use can present a problem. The noise of the Velcro being unfastened can startle a young or nervous horse and therefore care should be taken. If tapes are used, they should be fastened on the outside of the cannon bone.

FITTING A STOCKINETTE TAIL BANDAGE

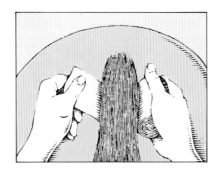

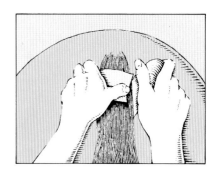

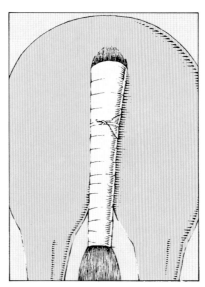

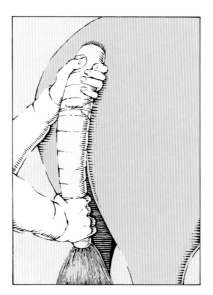

1 Dampen hair with water brush. Unroll short length of dry bandage and place this beneath the tail, close to dock.

2 Holding the end of the bandage against the tail, make one turn to secure the bandage. Then continue the bandaging evenly downward.

3 The tail bandage should stop just short of the last tailbone and the remaining length should be bandaged upward and secured with tapes.

4 Finally bend the tail into a comfortable position. Tail bandages should not be left overnight. Slide them off, downward, with both hands.

BELOW: *Four stages in fitting a stable bandage. Pad beneath all bandages with cotton or an equivalent. Wool stable bandages are rolled evenly down from below the knee or hock to the coronet, then upward to the start and tied on the side of leg.*

BELOW: *Placing of gauze exercise bandage to support back tendons and protect the leg. These bandages are applied firmly and often stitched in place for greater security.*

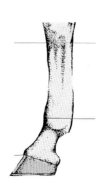
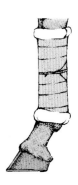

CLEANING TACK

RIGHT: *The necessary equipment. This consists of 1 Chamois leather; 2 Dandy brush; 3 Saddle soap; 4 Round sponge for soaping; 5 Flat sponge; 6 Bucket of tepid water; 7 Two cloths, one for polishing; 8 Metal polish; 9 Two stable rubbers.*

Ideally, saddlery should be cleaned each time it is used; it should, at least, be given a quick cleaning, with a thorough one once a week. If the leather is allowed to dry out, it will become brittle.

1 The lining of the saddle is washed with a damp sponge. Only leather-lined saddles should be washed like this.

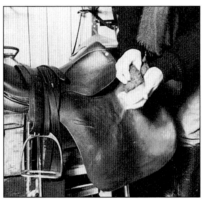

2 Applying saddle soap to the seat. Care must be taken not to overwet the leather, or water may seep into the stuffing.

3 Cleaning under the flap. This area attracts dirt and sweat, so it needs a thorough cleaning with a damp sponge.

4 Rubbing the seat and flaps with a damp sponge. The areas are dried with a chamois leather to remove surplus soap.

5 Washing the girth straps. These should be checked thoroughly for wear, since they are vital to the rider's safety.

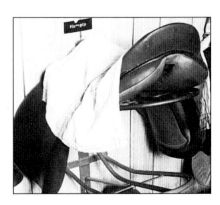

6 After polishing the saddle, it is replaced on its bracket or saddle horse and covered with a stable rubber.

CLEANING BRIDLES AND BITS

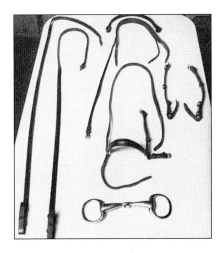

1 A dismantled bridle, consisting of reins, snaffle bit, noseband, headpiece, browband, and cheekpieces.

2 Rubbing the reins with a damp cloth. All leather parts should be cleaned similarly and then dried.

3 If the bridle is fitted with metal rings or studs, these should be cleaned with metal polish.

4 Washing the bit thoroughly to remove all traces of stains and saliva.

5 The rings of the bit – not the mouthpiece – should be polished with metal polish.

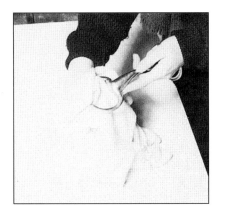

6 The final polishing. The bridle is then re-assembled and hung up ready for use.

7 Cleaning the girth. The method varies – webbing is brushed and scrubbed, nylon and string scrubbed.

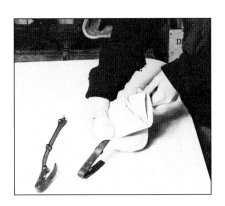

8 Cleaning the stirrup leather. This is done in the same way as the saddle, checks being made for wear.

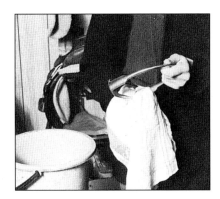

9 Cleaning the stirrup irons. After the dirt has been removed, the irons should be cleaned with metal polish.

RIDER'S CLOTHING AND EQUIPMENT

There is no need to buy all the correct riding kit for the first lessons. Riding clothes are expensive and it is advisable to make sure that you are going to pursue the sport before making such an investment. It is perfectly acceptable for beginners to dress informally.

Hard hat A hard hat is one essential item of riding equipment, and it should always be worn from the first time you sit on the back of a horse. The chin strap should always be secured, while, if you ever do fall on your head, the hat should be checked by a saddler to make sure the protective crown has not been damaged.

Raincoat Any well-fitting, waterproof jacket is suitable for riding. Make sure that it is not too tight around the arms, otherwise your movements will be restricted, but do not select one that is too baggy or voluminous. If you do, your instructor will not always be able to see if you are sitting correctly.

Turtleneck sweater or shirt Either of these items of clothing are perfectly acceptable. Choose the one which suits the weather best.

Jeans These should be tough and well-fitting. If they are too baggy around the legs, they will wrinkle against the saddle and lead to considerable discomfort. Stretchy slacks can also be worn, provided that they have a strip of elastic under the foot to stop them wrinkling up your legs.

Shoes These should be laced and be made of tough leather. They should never be fitted with any buckles or other adornment that could catch on the stirrup irons. The heels should be low and the sole should run through the entire length of the shoe – both the sole and the heel.

As soon as you are sure that you are going to ride seriously, it is

DRESSAGE CLOTHING

For Novice tests, a tweed or dark jacket is worn, white shirt, light-colored jodhpurs, stock or tie, long boots and an approved hat.

For Medium tests, dress is the same as for Novice tests but a stock is often worn instead of a tie.

At Advanced level, a black tailcoat and top hat are worn, or dark jacket and bowler. Top hats and bowlers should be worn straight on the head, resting just above the eyebrows. They should not sit on the back of the head.

worth buying at least some of the correct equipment. You will not only look more professional, but also feel more comfortable.

Hacking jacket This type of jacket is specially tailored for riding. The cut ensures you have sufficient room for free arm and shoulder movement; it also lets the coat fall smartly and properly down over the back of the saddle.

Shirt and tie Any shirt and tie is suitable, although it is conventional to avoid anything too loud, brightly colored, or patterned. Usually, it is acceptable to wear a turtleneck sweater under your jacket.

Jodhpurs These are trousers specially designed for riding. They are tight-fitting right down the legs to the ankles, which ensures your legs are not pinched by the stirrup leathers.

Jodhpur boots Jodhpur boots are ankle-length boots, which either have elastic sides or fasten with a strap. They are designed to be worn with jodhpurs (or with jeans), but not with breeches.

Gloves Experts disagree on whether riders should or should not always wear gloves. When it is cold, they certainly assure greater comfort. Remember, however, that they must be made of a nonslip material, such as string. Woolen ones are hopeless, as they will slip on the reins in the rain or if the reins are even slightly sweaty. Gloves should be well-fitting, and not too thick, as this reduces contact with the reins.

For formal occasions, such as hunting or a competition, a suitably formal outfit should be worn. There are a number of variations on this basic pattern, depending on the activity involved.

SHOW JUMPING CLOTHING

1 Correct dress depends on the level of the show.

2 For national and international classes, men usually wear red jackets. The chief exceptions are competitors from Eire who wear green.

3 For national and international classes, women usually wear a black jacket and either a shirt and collar, or shirt and tie.

Bowler hat A bowler is slightly more formal than a hard hat, but it is designed to fulfill exactly the same purpose. It is worn by both men and women.

Black jacket This is designed and cut in a similar fashion to the tweed hacking jacket, but, again, it is somewhat smarter.

Stock This is a specially shaped "cravat," which is usually made of white material and is worn with a collarless shirt. It has to be tied in a special way and is normally secured with a plain gold stock pin (like a stick pin).

Gloves On formal occasions, clean, string gloves should be worn.

Breeches Like jodhpurs, breeches are designed specifically for riding. They are always worn with long boots, so breeches do not necessarily extend to the ankle.

Riding boots Riding boots, unlike jodhpur boots, come up to the knee. At one time they were always made of leather, but, today, special "rubber" riding boots are available. These are perfectly

TYING A CRAVAT

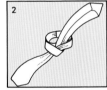
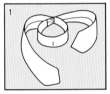
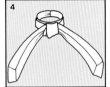
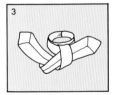

The stock is a specially shaped cravat, usually made of white material and worn with a collarless shirt for formal occasions. It has to be tied in a special way, which may seem complicated at first, but will soon become second nature.

adequate for most occasions and have the advantage of being considerably cheaper than their leather counterparts.

Sticks and spurs A rider formally dressed almost always carries a riding cane and may wear spurs. Canes and whips should be rigid and not too long or "whiplike." Spurs should always be blunt; the best are made of stainless steel.

It is a good precaution to wear a comfortable body protector under your jersey.

CHAPTER 3

LEARNING
TO RIDE

FIRST LESSONS

In the early stages of your riding career, try to establish and maintain a regular pattern of lessons. If you can only manage to ride once a week, book a course of six half-hour lessons. This is usually sufficient to give you the initial feel of riding and after these you can move on to one-hour lessons.

The more frequently you have lessons, the more quickly your riding will progress; it is far better to have a half-hour lesson each week than a one-hour lesson once a fortnight. Moreover, you will find it far too tiring to try to ride for a full hour at a time at first. If you have been working properly for the first few weeks, your muscles will be aching and your legs will feel as if they are ready to drop off after considerably less than a half-hour. Much of the technique of riding depends on developing the correct muscles and, until you have done so, you cannot hope to begin to realize your potential as a rider.

Mounting the horse

Once in the school, your nervousness or apprehension may well heighten with the thought of actually getting on the horse. There is no need whatsoever to worry about this, provided that you think logically about what you are doing.

The first step is to take up the reins. Even though the assistant will be holding the horse's head, you should get into the habit of holding the reins sufficiently tightly yourself to discourage the horse from moving forward. Take them up in your left hand – together with a good lump of mane if you feel the need for security – and stand by the horse's shoulder facing towards his tail. Take hold of the stirrup iron with your right hand, turn it toward you and put your left foot into it. Hop forward and turn to face the saddle. Then, with your right hand across the seat

of the saddle, spring up off your right foot, trying to avoid hitting the horse's back as you do so, and sit down lightly in the saddle. Put your right foot in the stirrup iron.

The stirrup leathers should lie flat against your legs; if they are twisted, one edge will dig into your legs and this will soon be very painful. You must adjust the stirrup leathers too, so they are the right length. It is impossible to give exact instructions as to how to judge the correct length, as this will alter as you settle deeper into the saddle and become more confident. As a guide – always providing that this length does, in fact, feel comfortable – hang your legs straight down and adjust the leathers so the bottom of the iron is level with your anklebone. The leathers must be of equal length; if your legs are lopsided, you will have no hope of sitting straight and being equally balanced in the saddle.

LEG POSITIONS

ABOVE: *Your lower leg should be in contact with the horse's side all the time. To apply the leg, you close your lower leg and ankle against the horse's side, still using the inside of the leg, not the back of the heel.*

ABOVE: *The inside lower leg is applied on the girth to create impulsion and to instruct the horse to bend around that leg when making turns and circles.*

ABOVE: *The outside lower leg is applied behind the girth, to make sure that the hindquarters do not swing out but follow the line of the front end.*

ABOVE: *The stirrup iron is under the ball of the foot, and the heel is lower than the toe. This makes it easier to apply the legs correctly. You may have difficulty at first in keeping the heel below the toe.*

Position in the saddle

There are two things to be said straightaway about adopting the correct position in the saddle. The first is that it is of paramount importance, for it is only by sitting correctly at all times that you can become an effective and good rider, and the second is that you should not expect it to feel comfortable until you have got used to it.

If you are sitting up sufficiently straight with your head held high, it will feel as if you are sitting as straight as a ramrod, with your head and shoulders forced back. If you could see yourself, you would see that this was not actually so; in any event, the importance of keeping your head up cannot be sufficiently stressed. Your head is the heaviest single part of your body and its position determines much of the positioning of the rest of it. If you look down, your weight automatically shifts forward. A horse already carries two-thirds of his weight on his forehand, so the last thing he wants is your extra weight on this area.

Your body weight should be on your seat bones, evenly distributed on either side of the saddle, so that you and the horse can balance freely. Your seat is maintained at all times by balance, not by grip. If you grip with your thighs and knees, you are automatically impeding the horse's freedom of movement and pushing your seat upward in the saddle. Try to relax your knees and hips, which will demand some conscious thought and effort.

Your legs should hang straight underneath you, making a straight line from the shoulder and hip to the heel. This should not be done by bending more acutely at the knee, so the lower part of your leg is forced back, but by bringing that part of the leg directly underneath you. Finally the weight of your legs should fall into the heels. In doing this, the temptation is often to stick your toes out, which results in gripping the horse's sides with the back of your leg. Instead of thinking of pressing your heel down, think of bringing your toe up, pointing directly forward.

COMMON FAULTS

1 If the rider's hands are uneven, the bit will not lie correctly in the horse's mouth, and the tension on its mouth will be wrongly distributed.

2 The horse is reacting to a stiff arm and unyielding hand by tilting its head, grinding its bit, putting its ears back and refusing to walk on actively. The rider's stiffness is causing the problem. This can be solved by working on developing a secure, independent seat.

3 Do not ride with your lower leg sticking out from the horse. If you try to apply the leg aids from this position, your horse will have a bad fright.

The first movements

Even before you learn to hold the reins, your instructor will probably ask the assistant to lead the horse forward, so that you can get the feel of the walk. With your hands on the pommel, close the insides of your legs against the horse's side, and by doing this you will encourage the horse to move forward.

A horse walks by picking up and putting down each foot separately, so that four distinct hoofbeats can be heard. This results in a pace that has a gently swinging or swaying movement and, as your aim at all times is to move with the horse, you should let your body sway gently in time with the movement. This is not a conscious movement. If you think about it, you are likely to sway too much, so concentrate instead on maintaining the correct position in the saddle. Keep sitting on your seat bones, with your back straight and your head held up so you look between the horse's ears, and your legs hanging easily beneath you.

Even though you are now in the saddle, you will either be led by an assistant or longed by your instructor. This means he or she will

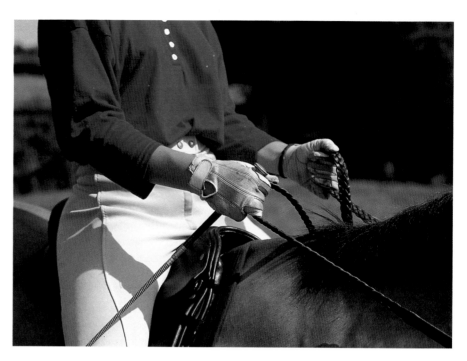

ABOVE: *The horse's mouth is very sensitive, and your hands must be sympathetic if it is not to become hardened or damaged. You hold the reins by wrapping the fingers around the rein, almost closing the hands to make a fist. This gives you a firm grip on the reins without having to use your arms. The arms remain soft and relaxed. The horse's movement is contained by putting pressure on the bit through the reins. The severity of the aid will depend, to some extent, on the sensitivity of the horse. You must match what the horse offers you.*

stand at a given spot – the center of a circle – and control the horse from the end of a long rein attached to his bit, guiding him around and around. In either case, you do not yet have to worry about controlling the horse; you can concentrate completely on sitting correctly. Until this becomes second nature, this will take all concentration.

Holding the reins

Although you do not need to concern yourself with controlling the horse yet, you must know how to hold the reins. Pick up a rein in either hand, so that they run through your little and third fingers, across the palms of your hands and emerge between your

ABOVE: *The rider is using her inside leg to create good forward movement and at the same time is maintaining a steady contact with the horse's mouth. It knows it is not going to be jabbed in the mouth by the bit, so it is going forward confidently into the rider's controlling hand.*

ABOVE: *The rider drops the reins so that the horse can stretch out as it walks. This is a good way to reward a horse and allow it to relax when it has been working strongly in walk and trot.*

thumb and first finger. Your thumbs – pointing straightforward – are on the tops of the reins holding them down onto your first finger.

Hold your hands in front of you about 4in (10cm) above the withers with elbows bent and supple. There should be a straight line from the bit, through the reins, hands, and arms up to your elbow. You will find this necessitates holding your hands quite far forward and apart about the width of the bit. Think of holding a book in front of you, your thumbs on top of the pages.

Always remember that the reins are for guiding the horse and not to be clutched for security, to maintain your position, or to correct your balance. Think of them as delicate threads: if you pull on them too hard, they will break.

As soon as the horse moves forward – remember you make him do so at this stage by closing the inside of your leg against his sides – you must make sure you follow the movement of his head with your hands. If you keep your hands rigid, you will naturally impede his head movement, thus making it difficult for him to balance properly. To do this, he needs to be able to move his head freely. You will find that, at a walk, his head moves quite considerably, so let your hands move as well. Again, do not try to move them consciously; just let them follow the movement, so it is the horse's head that moves your hands rather than the other way around.

When you want the horse to slow down or stop, you must pass on your intention to him. Clearly you have

to discourage or restrict his forward movement; you do this by pushing him forward with your legs and then squeezing the reins gently so that he meets resistance. In other words, your legs push him up to meet your hands. Do not tug on the reins, leaning backward as you do so; instead, brace your back muscles, so you are no longer following the horse's movement, and close your hands on the reins, so that they, too, are no longer following the movement. This should be done gently, as though you were squeezing a sponge.

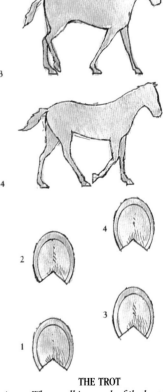

THE TROT
ABOVE: *When walking, each of the horse's feet lift in sequence: (1) near hind; (2) near fore; (3) off hind; (4) off fore: a four-time pace.*

COMMON FAULTS

1   The rider has let her reins go, losing all contact with the horse's mouth. She is slouching in the saddle and is not pushing the horse on with her leg. As a result the horse is ambling along with his head dropped, rather than walking out with even steps.

2   The rider's arms are straight and stiff, her hands are too low, and they are not giving with the horse's movement. The horse is fighting against this restriction by raising its head and tipping it to one side.

3   The rider has crept up the horse's neck so that her weight is a long way out of place, and she is holding the horse on a very short rein. This makes the horse anxious and it starts to jog.

From walk to trot

If you feel reasonably confident and comfortable at a walk in your first lesson, your instructor may suggest that you try a few strides at the trot. This is a very different pace from the walk. It is a two-time pace – that is, the horse moves its opposite diagonal legs together and springs from one pair of diagonals to the other. This makes it a much bouncier gait, so prepare to be bounced around in the saddle.

As you progress, you will learn to ride at this pace, both by sitting down in the saddle to the movement and rising up and down with it. For this first time, however, take hold of the pommel firmly with both hands, ask the horse to move into a trot by closing the inside of your legs against his sides and then try to maintain your balance as he moves forward. When you want him to return to a walk, try to sit very tall in the saddle – this helps you to maintain your balance as the horse slows down – and ask him to walk by closing your legs against his side and bracing your back muscles.

The rising trot

By now, you should be confident enough in the saddle to learn how to rise at the trot. From your previous experience, you will know that the trot is a very bumpy pace, so, to avoid any unnecessary damage to the horse's mouth, take hold of the neck strap before moving.

At first, it helps to practice the technique of the rising trot while the assistant holds the horse still. Gently stand up and sit down again in the saddle, using your knees as hinges to push you up rather than relying on the stirrup irons to support your weight. Thinking of this action as a forward and backward movement from your hips helps resist the temptation of rising too high. As you rise forward and

A GOOD TROT

ABOVE: *The rider uses her legs to create impulsion and the horse steps out well. Its hind leg comes well underneath its body,* *showing that its hindquarters are engaged and working actively.*

COMMON FAULTS

1   The rider is behind the horse's movement. Her body is no longer upright, her arms have become stiff and she is balancing by hanging onto the reins. In response the horse has raised its head and neck and indicates that it is not comfortable with the rider's position.

2   The rider has let her reins go long and is sitting too far forward. The energy she is creating by using her legs, is wasted as she has lost contact with the horse's mouth and therefore cannot contain the horse's movement. As a result, the horse's weight is falling onto its forehand.

THE TROT

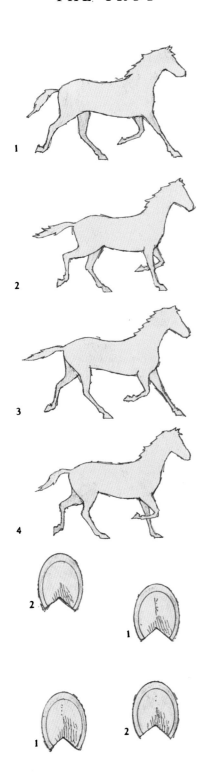

ABOVE: *In trot the sequence of footfalls is: (1) and (2) off fore and near hind together; (3) and (4) near fore and off hind together: a two-time pace.*

backward count "one, two," keeping the counting and the rising as regular as possible.

This may seem easy, but transfering it to the actual action is more difficult, although it soon comes with practice. Therefore, if you find you are rising in time with the movement for a few strides and bumping for double the number, you can be pleased, for it shows you are making progress. In these early stages, only trot for short distances at a time – down one side of the school, for example. The trot is a very tiring pace for you until you get used to it: if you get overtired, you will find it even more difficult.

Dismounting

The last stage of any lesson is dismounting. To do this, take both feet out of the stirrup irons and collect the reins in your left hand. Put your right hand on the pommel, or the side of the horse's neck, and swing your right leg up behind you and over the back of the saddle. As gently as you can, slide yourself down to the ground, letting your knees bend to take the jar as you land. Be warned – after even a half-hour on a horse, it takes a second or two to regain your usual muscles, and you will probably find your first few steps are somewhat staggering ones. It is not unknown for people to find themselves sitting down on the ground.

Thank your horse for the ride by patting him. Then, with the reins looped around your arm so that he cannot walk off on his own, run the stirrup irons up the side of the leather closest to the saddle so they rest against the top sides of the saddle. This will prevent them from banging against the horse's sides as he walks. Take the reins over the horse's head and, holding them together with the leading rein, lead him away.

DISMOUNTING

1 Leaning forward slightly, take both of your feet out of the irons.

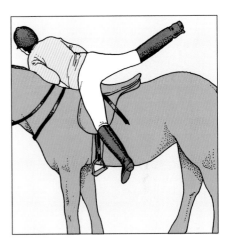

2 Leaning forward with the left shoulder, swing right leg up and behind.

3 Still holding onto the saddle, bend your knees as you land, to prevent jarring.

The sitting trot

Once you can maintain the rising trot, you can begin to learn the sitting trot. For this, as the name suggests, you sit in the saddle all the time; this is far more difficult than it sounds, particularly if your horse has a very springy action. Try sitting for a few strides and then going back to a rising trot, gradually increasing the amount of time spent sitting as you begin to feel more comfortable. Relax your body, sitting deep into the saddle, but sit even more upright than for the rising trot. The tendency is to become careless in an attempt to relax into the movement, but you will find, in fact, that this throws you out of the saddle even more vigorously.

The sitting trot is a tiring pace for horse and rider, particularly when you are learning it. Therefore do it only for short bursts at a time. Rise to the trot down the long sides of the school and sit along the shorter ones. Remember with all trotting exercises – both sitting and rising – to work evenly on both reins.

Positional exercises

It was briefly mentioned earlier that the position in the saddle should be maintained by balance at all times.

This is very important; if you grip with your thighs and knees, you will automatically make the horse tense his back and shoulder muscles. This, in turn, restricts the freedom of his movements. It also pushes you upward in the saddle so you come off your seat bones. To help you achieve an even balance across the horse's back, let your arms hang down by your sides. Then, without rounding your shoulders, imagine you are carrying two heavy shopping bags – one in either hand – which, by definition, require you to be evenly balanced.

To maintain a balanced position, all parts of your body must be perfectly relaxed and supple. Even though your back is straight and your shoulders are pulled squarely back, they should both still be relaxed and so should your hips and pelvis. Your knee, hip, and elbow joints should all be supple. It is essential to coordinate body movements – that is, moving one part of the body in harmony with another – but you must also be able to move one part of your body independently of the other parts. To begin with, you will probably find this extremely hard to do; as you use your legs, your hands will automatically jerk upward or make you lean backward, but, as you think about it and practice it, you will find it becomes easier.

Every lesson or schooling session should include some exercises, for they are invaluable in developing the position in the saddle. They will help you to coordinate your body movements as well as use each part independently; they will help you to become supple and reduce stiffness; and they will help to develop the correct riding muscles.

Some of the exercises pictured here are designed to help you learn to move one part of your body independently. Provided that the assistant is still leading you, you can do them at a walk as well as at a halt.

Riding without stirrups

From an early stage, you should get accustomed to riding without stirrups. This will help you with your balance; it also teaches you not to rely on your stirrup irons to help you maintain your position.

With the horse halted, take both feet out of the stirrups and cross the leathers over in front of the saddle, the irons resting on either side of the horse's neck. The usual order from your instructor for this will be to quit and cross your stirrups. Ask the horse to walk on in the normal way and let your legs hang down in a

POSITIONAL EXERCISES

1 Cross arms and bend forward from the waist keeping your lower body still. Slowly straighten up and lean backward the same way, making sure your legs do not shoot forward. This will help supple your waist.

2 Put one hand on the pommel and circle the other arm from the shoulder. Think of trying to pull your hips out of their sockets as you lift your arm upward. Keep your legs absolutely still and in the correct position. Repeat with the other arm.

3 With both arms outstretched at shoulder level, turn from the waist first to the right and then the left, so that your hands point to the horse's ears and tail.

relaxed fashion for a few paces. Put your legs in the correct position, then raise your toes, letting the weight fall down to the heel again. Keep your hands on the pommel initially, but when you feel secure, take up the reins as you would normally. You must be even more careful now not to rely on your reins to balance you.

A good exercise to do while riding without stirrups is to swing your legs from the knee backward and forward, either together or alternately. Alternatively, try swinging one leg forward and the other back at the same time, making sure you keep your hands still and do not bounce in the saddle. This helps to supple your knees.

Always remember that, as the majority of exercises are designed to develop muscles, always to repeat those that involve just one side with the other one. By the same token, it is very important to ride evenly at the same pace on both reins, otherwise both you and the horse will develop a preference for riding one way. Largely because they are ridden by so many people, most riding-school horses do have a preference for one side and you will soon feel this: they will always be a little stiff and more reluctant to cooperate on the other rein.

EASY EXERCISES WITHOUT STIRRUPS

1 Swing your legs backward and forward from the knee.

2 Keeping legs still, circle each arm from the shoulder.

COMMON FAULTS

1 Hanging onto the reins to maintain balance.

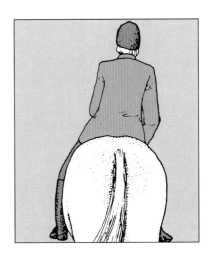

2 Moving out of position while doing exercises.

COMING TO A HALT

With a young horse, aim to bring the horse up with front legs square, and hind legs nearly square.

COMMON FAULTS

1 The rider has shifted forward. The horse has its weight on its forehand and its hindquarters are unevenly balanced. The rider was in the wrong position to communicate with the horse.

2 The rider is leaning back and pulling on the reins to bring the horse down a pace. The horse becomes resistant to the rider's action, so the pace following the transition will be affected.

The art of control

Now that you are gaining strength, confidence, and a sense of balance, think more about controlling your mount. The time has come when the position you have worked so hard at has also to be effective in making the horse obey and work for you. Combining a correct position with effectiveness is often one of the hardest aspects of learning to ride.

The signals that have been devised to help a rider transmit his wishes to his mount are known as the aids. Natural aids are those given with parts of the rider's body. The chief ones are the hands and the legs; in addition, shifting the body weight slightly or bracing muscles are included. The voice also counts as a natural aid, though this should be used sparingly and never in anger. Artificial aids are the additional items of equipment a rider can use to help to encourage or control the horse. These consist of a cane or whip, spurs and ancillary items of tack, such as a martingale, which are used for specific reasons to do with control.

The most important point to remember about aids is that they should always be given firmly and decisively, but never roughly. A half-hearted aid is useless and the horse will not understand what he is meant to do. Aids should be given together, that is, a leg aid should be supported by a hand aid.

If, for example, you want your horse to move forward more quickly or alertly, you would tell him by squeezing his sides with the inside of your leg. At the same time you must relax or yield with your hands in order to give him the freedom to move forward. Similarly, if you want him to slow down, you push him forward with your legs so that he gathers himself together, then stop the forward movement by resisting with your hands and bracing your back muscles.

At this time, you should learn to carry a whip during lessons. Choose either a cane or a short, fairly rigid

ABOVE: *The rider's position is good for this trotting exercise. Her weight is slightly forward, her hands are in contact with the horse's mouth, she is sitting down in the saddle, and squeezing the horse forward with her lower legs.*

ABOVE: *Here the rider is cantering over poles and she has folded forward, closing the angles. Her lower legs are on the girth and are maintaining a good contact with the horse. Her body is forward and, as she is not leaning on her hands, they are in active contact with the horse's mouth. She is driving the horse on with her lower legs.*

riding crop. Hold it in your hand along the rein so that the top emerges from the crook of your thumb and forefinger. The remainder of the whip rests across your lower thigh.

When riding in a school, a whip should generally be held in the inside hand. This is because a horse

will automatically move away from it when it is used and he should move always toward the outside of the school, not in toward the center. If you are carrying a whip, therefore, you must remember to change it to the other hand when you change the rein. Do this as you cross the center of the school.

COMMON FAULTS: WALK TO TROT

1 The rider's legs are away from the horse's sides, completely out of contact with the horse, so the instruction to change pace surprises it. The rider has let the reins go loose as well. The horse reacts by hollowing its outline and shortening its steps.

2 The rider has anticipated the transition by tipping forwards and dropping contact with the reins. As a result there is no controlling hand for the horse to go forwards into. It jumps into the next pace rather than stepping forwards into it.

3 The rider has failed to achieve a good, active pace before the transition, and is also leaning forwards. As a result, the transition is sluggish.

4 The rider has let the reins go loose, losing contact with the horse's mouth, and she is not creating impulsion in the horse's hindquarters. The horse's weight is on its forehand, and the transition is poor.

From trot to canter

COMMON FAULTS: TROT TO CANTER

Once you have truly mastered the sitting and rising trot, the next pace upward is the canter. This is executed by the horse in quite a different way from a walk or trot and has a very different feel from the paces so far discussed. It is a pace in three-time in which one hind leg strikes off, to be followed by the diagonal of the opposite hind leg and foreleg, and then by the opposite foreleg. There follows a moment of suspension when all four legs leave the ground before the pattern is repeated.

To go from trot to canter, sit deep in the saddle, close your hands slightly on the reins to prevent the horse from going into a faster trot, and press both legs against his sides. Lean forward slightly from the waist to counter the horse's movement, but this should be barely perceptible.

As the horse breaks into a canter, try to sit deep in the saddle. Make

1 The rider is holding the horse on too tight a rein, and her hands are not allowing the movement of the horse's head. The horse is fighting against this restriction by raising its head and neck and resisting the movement.

2 The rider is standing up in the saddle and leaning forward. From this position she cannot push the horse forward into a canter.

no conscious effort to move; let the rhythm of the pace move you. Avoid using your hands to balance as, at the canter, more than any other pace, you need to let your hands move with the horse to give his head the freedom it needs.

To return to the trot, slow the horse down by pushing forward with your legs, then resisting with your

hands and bracing your back muscles. Then try to pick up the rising rhythm as soon as he begins to trot.

Only attempt the canter for a few strides at the beginning. Concentrate on relaxing into the saddle, then performing smooth transitions from a trot to a canter and a canter back to a trot.

TURNS AND CIRCLES

A recognized riding school measures 40 × 20ft (12 × 6m) and is usually marked with a standard series of letters. So far you have concentrated on riding around the outside and changing the rein across the center of the school or from the three-quarter markers, but there are other ways in which you can use the school.

The first school figure to practice riding – at a walk, then a trot and finally a canter – is a 20ft (6m) circle. The command from your instructor to ride a 20ft (6m) circle and then to return to the outside track will be: "At C (or A) go forward into a 20ft (6m) circle. As you return to C (or A), go large." Leave the track at the center of the short side of the school (C or A) and ride in a perfect arc to a point between the three-quarter points and the center of the school – between H and E, K and E, or M

BELOW: *The aim, when turning through a bend, is to keep the horse looking in the direction in which it is moving. Its neck should have no more bend in it than the body.*

and B and F and B according to which way you are going. Then go on to X (the center of the school) and next to a point between the other three-quarter marker and the center of the school. Finally, go back to C or A.

Riding a perfect circle is not easy, so concentrate and make sure your horse is properly bent around your inside leg all the time. Bear in mind that his body should be bending around the arc of the circle – at no point are you attempting to turn him. Watch that he does not "fall in" to the center of the circle so the circle becomes imperfect. It is helpful to ride on soft ground, as you can check hoofprints to measure the accuracy of the circle.

When you have ridden a few

circles, change the rein and ride in the opposite direction, so neither you nor the horse do not favor one side. Also, try riding smaller circles, remembering that your aim is to form the horse's body into a perfect arc throughout the exercise.

The serpentine tests your control and ability to ride accurately to the full. This involves riding down the school from C to A, making four perfect loops, the extreme point of which is about 10ft (3m) from the side of the school. The whole movement should be smooth and fluid, not jerky or uneven. Try the exercise at a walk first, and then progress to a trot, checking your tracks in the school to see how accurate you have been each time.

If you are riding with others in

LEFT: *The rider closes the inside leg against the horse to ask it to move forward away from the leg while using the outside leg behind the girth to prevent the horse's hindquarters swinging out. At the same time she uses the inside hand to ask the horse to turn and controls the degree of bend with the outside hand. She looks in the direction in which she wants the horse to travel. She is applying the aids in a well coordinated way to produce a good turn: the horse's head is just inclined in the direction of the movement and the hind legs are following in the path of the front legs.*

RIDING CIRCLES AND TURNS

ABOVE: *When riding a circle, imagine a diamond shape on the ground. Ride around the diamond, rounding off each point. This should give you a correct circle. Turns consist of a section of a circle, and are ridden in the same way.*

RIDING LOOPS

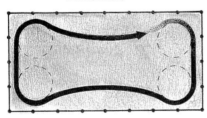

ABOVE: *Shallow loops made on the long side of the area will teach the horse to change the direction of the bend and make it more supple. You can introduce a small circle at the end of the loop.*

SERPENTINES

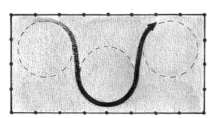

ABOVE: *Once you are progressing well on circles and loops, you can also introduce serpentines into the routine, across the width of the area.*

the school, there are several exercises and movements you can perform together, which will help to improve your skill, control, and accuracy. Such maneuvers include, for example, being told by your instructor to position yourself at a letter and then changing places with another rider standing at a different letter. If there are several of you in the school at one time, this maneuver can be quite a test of your riding. The code in passing other riders head-on is to pass right-hand to right-hand.

In addition, practice riding holding both reins in one hand, so that this will not present a problem to you, if it should ever be necessary on a ride. If you are holding the reins in your left hand, for example,

hold the left rein normally, then bring the right one across the top, so that it enters your hand between the thumb and forefinger, crosses the palm, and emerges beneath your little finger. When riding like this, you must control the horse more actively with your legs, bending it around your inside leg if you want to turn corners. Do not attempt to guide it too accurately with your hand – this is bound to confuse.

To help you think less actively about the hand holding the reins, put your reins into one hand in order to allow you to do something else with your other hand – such as blowing your nose, doing up a button on your coat, or adjusting the length of a stirrup leather.

COMMON FAULTS

1 The rider is trying to make the horse bend into the corner by holding it out with her outside hand, instead of using her legs to control the movement. As a result, the horse is looking in the wrong direction and has hollowed against the rider.

2 The horse is bending its neck too much to the inside of the circle. If the correct degree of bending is being maintained, the rider should be able to see the corner of the horse's eye, but this horse has far more of its head turned in.

3 The rider is looking too far around the circle, so her head and upper body are turned too much to the inside of the circle. The rider should be looking through her horse's ears, with her shoulders and hips parallel to those of the horse.

LEARNING TO JUMP

Although the aim is always to produce one continuous flowing movement, the horse's jumping action can be broken down into five elements – approach, takeoff, suspension, landing, and recovery. During the approach, the horse, having seen and summed up the obstacle in front of him, will balance and prepare himself for the jump by stretching his neck and lowering his head. He may begin to lengthen his stride, but he should continue at the same even pace, without altering his speed.

At the point of takeoff, the horse brings his head up as he lifts his forehand off the ground. The power for the leap forward comes from the horse's hocks, which are tucked well beneath him to act like a spring. During the moment of suspension, the horse's body forms an arc over the jump with the head and neck stretched forward. As the descent begins, the horse extends his forelegs, his head and neck down towards the ground and tucks his hindlegs under him, so that they will clear the jump. As the forelegs touch the ground – usually one just in front of the other – the horse balances himself by bringing his head up and shortening his neck. The hind feet touch down immediately behind the forefeet, one forefoot often moving into the next stride before the hind feet land.

At all times the rider must remain in complete harmony with the horse, taking particular care not to interfere with the free movement of the animal's head. The jumping position described in the following section has been designed to fulfill this aim.

If jumping is to be included in a lesson, it should come toward the end. This will give the horse time to loosen and limber up and make sure that the rider is sitting deep in the saddle and loosened up enough to be riding at his or her best.

ABOVE: *The shortened leathers have the effect of moving your seat toward the back of the saddle so that the lower leg and knee can sit firmly around the horse. Your seat comes slightly out of the saddle and your body folds forward from the hips. Keep your back straight and look ahead. Your elbows remain bent and your hands must follow the movement of the horse's head.*

BELOW: *You must keep your weight over the horse's center of gravity by folding forward over a jump. The shoulder, elbow, hip, and ankle remain in a straight line.*

The jumping position

The first thing to learn and practice, until it becomes second nature, is the jumping position. Assuming that you will have dropped your stirrup leathers a hole or two for general school work by now, you will need to take them back up again for jumping. The stirrup leather should still remain vertical to the ground, which means that your knee and ankle will be bent a little more deeply. This allows them to do their work as "hinges" and "shock absorbers."

In the jumping position, the rider's upper body bends forward in a folding movement from the waist. The back remains straight and supple – there should be no slouching or rounding of the shoulders – while the head is still held high, looking straight between the horse's ears and never down at the jump. The seat should remain in light contact with the saddle throughout the approach, although it may lift up during takeoff. The body weight is taken on the knee, thigh, and heel, but resist any temptation to straighten the knee

CLOSING DOWN THE ANGLES

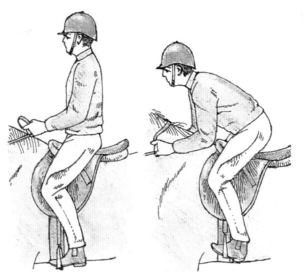

The stirrup leathers are shortened for jumping, closing down the angles at hip, knee, and ankle. Think of your body as being a "W" turned on its side: shoulder to seat, seat to knee, knee to heel to toe. In the correct jumping positon, you flatten the W as much as possible.

and to stand up in the stirrup irons. There should still be a straight line from your elbows, which are bent and remain close to your sides, through your arms and hands, along the reins to the bit. Your shoulders, elbows, and fingers have to be even more supple in order to follow the movement of the horse's neck. In fact, your hands should stay in the usual position throughout a jump and it is better to move them forward rather than run any risk of jabbing the horse in the mouth.

In the early stages of jumping, make sure a neck strap is buckled around the horse's neck and hold onto this. It will make you feel more secure as well as guarding against you jerking the horse's mouth. Practice moving into the jumping position in the saddle, first at a halt and then at a walk and trot. When you can bend forward and sit up straight again without losing your balance at a trot, try doing the same exercise at a canter. Your aim is to achieve a smooth rhythm.

COMMON FAULTS

1 The rider's seat has come too far out of the saddle, causing her legs to straighten and push forward and her back to round. Her weight is too far back, and she will get behind the horse's movement over the fence.

2 The rider has folded too far forward, causing her to hollow her back. It is impossible for her to keep her leg in the correct position. The lower leg has moved too far back behind it. She cannot communicate effectively with the horse from this position.

3 By standing in her stirrups the rider has opened up the angles at hip, knee, and ankle, and raised her hands too high, so that her position in the saddle is made very insecure.

Pole work

This is a very good exercise for developing rhythm when teaching a horse to jump, or when improving the performance of an older horse. It is also excellent for practicing the correct jumping position.

It is important that you have the distances between the poles correct. A horse will take two strides at the trot to one at the canter. If you have the poles double-spaced, that is, correct for the canter, they will be correct for the trot as well, and you will not need to keep dismounting to move them each time you change pace.

The distance between the poles will depend on your horse's size and stride. The table gives a guide to distances.

When the horse is happy and at ease going over one pole, move up to three poles or more. Do not work over just two poles, as it might encourage the horse to jump both poles together.

ABOVE: *The rider's position is good for this trotting exercise. Her weight is slightly forward, her hands are in contact with the horse's mouth, she is sitting down in the saddle, and squeezing the horse forward with her lower legs.*

Begin by walking over the poles, and then go over them at a rising trot. Cantering to poles should not be attempted unless the horse has mastered them at the trot. In canter, the horse should just bounce along without taking any steps between the poles.

COMMON FAULTS: TROTTING

1 The rider's leg is too far forward and has straightened, and her back has rounded. This often happens when the horse rushes the poles.

2 The rider approaches the poles with her legs too far back along the horse's side, and using the reins to balance. She cannot give the horse precise instructions from this position, and the horse is showing its concern at the lack of communication from the rider by raising its head and hollowing its outline.

3 The rider is resting her weight on her hands, and is not making use of the horse's impulsion, which she is trying to create with her legs. The horse is not taking a full stride as it approaches the poles.

Distances for pole work		
size of horse	length of stride trot	length of stride canter
14½ hh	4 ft 1.2 m	8 ft 2.4 m
16 hh	4½ ft 1.4 m	9 ft 2.75 m

ABOVE: *Here the rider is cantering over poles and she has folded forward, closing the angles. Her lower legs are on the girth, and are maintaining a good contact with the horse.*

Her body is forward and, as she is not leaning on her hands, they are in active contact with the horse's mouth. She is driving the horse on with her lower legs.

COMMON FAULTS: CANTERING

1 Although the rider's leg is in the correct position, her body is collapsing forward and to one side. She is leaning on her hands and looking down. As a result, the horse's movement is restricted.

2 The rider is sitting too upright, which puts her weight too far behind the horse's movement. The horse's balance and judgment are adversely affected, and it does not stride cleanly over the poles. Instead, its forefeet are either side of the pole.

The first jump

The next step is to hop over a small jump, which should be no more than about 10in (25cm) high; you can use proper jump supports or improvise by using stout poles on wooden boxes or barrels. Most riding establishments, however, will probably use a type of pole known as a cavalletti. If you do not use cavalletti, make sure the poles are thick and solid with no rough parts or sharp nails protruding. Horses show far more respect and jump better and more boldly over solid objects, rather than flimsy, unimposing ones.

Even though the horse is quite capable of stepping over the cavalletti at a trot, he will probably prefer to hop over it, so make sure you are prepared. The first "jump" you take is bound to throw you off balance; remember to hold onto the neck strap, so that you do not jerk the horse's mouth by mistake. When you are reasonably confident and able to maintain your balance and rhythm at the trot, approach the cavalletti at a canter. Keep the pace calm and let the horse bounce over the jump, offering no interference so he can take it in his stride.

After this, try placing another cavalletti or low jump, perhaps slightly higher, several strides further on, so you have time to return to the normal position before resuming the jumping position for the next one. If you have approached the first one calmly and quietly, there should be no need to check the horse between jumps. It is better not to interfere with the reins

TAKING THE POLES AND JUMP

1 The horse is approaching on a good stride and is looking intently at the fence. The rider is keeping a good conversation going with the horse via her lower legs.

2 The rider is closing down the angles at her hip, knee and ankle as she prepares for the moment of takeoff. The horse is reacting to her positive instructions by coming into the jump with impulsion and looking alert.

3 The rider has folded her body down well and is looking straight ahead. The horse has its weight over its hocks and is beginning to lift its forelegs.

4 The horse take the fence in its stride.

between jumps as both of you will probably lose your balance.

Having achieved good rhythm and balance over a couple of small jumps positioned some distance from each other, bring them closer together, so that they are separated only by a couple of strides. Again, allow your horse to judge the takeoff points and distance between the jumps: you should concentrate yourself on keeping your balance and not interfering with his movement in any way.

1 The rider is standing in her stirrups, straightening her body. She is keeping her balance by bracing her hands against the horse's neck. Her whole position is weak and insecure. If the horse chose to swerve out or stop, she would find it very difficult to counteract such misbehavior.

COMMON FAULTS

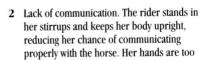

2 Lack of communication. The rider stands in her stirrups and keeps her body upright, reducing her chance of communicating properly with the horse. Her hands are too high, and the horse is on too long a rein. The rider's lack of positive signals is causing the horse to lose impulsion. Its back feet are trailing behind and it does not look committed to jump. The horse tackles the jump, but its ears are back, its outline is hollow, and it is taking off unevenly from its hocks.

Individual fences

As you gain confidence through working on grids of poles with a small fence, you can move onto tackle larger, individual fences.

When practicing over larger fences you should retain the last pole on the ground in front of the fence to act as a placing pole. It will bring the horse to the fence in the correct spot for takeoff.

In order to approach a fence correctly you need to concentrate on establishing a good, rhythmic canter with plenty of impulsion, and should wait for the fence to come to you. As with the previous exercise, it is important that you maintain the same rhythmic stride throughout both the approach and the jump itself. You should not need to check the horse at all as you approach the fence.

Present the horse to the fence with your legs wrapped around it. Try to have your lower calf and ankle in contact with the horse's sides, applying a steady pressure to tell the horse that you want it to continue. If the horse is confident that you will allow its movement with your hands, it will jump without your needing to force or yank it into the air. Keep your body relaxed and stay in balance with the horse over the jump.

If the horse does not give a good jump, it could be for a variety of reasons. The placing pole could be at the wrong distance from the jump, bringing the horse in incorrectly, so check that you have it in the right place. It could be that the horse has had a bad experience in the past. Or it could be that the horse lacks confidence in its rider. As with other riding problems, always ask yourself if you are doing everything correctly.

Above: *The rider is balanced well over the horse. She has folded her body right down. Her lower leg and ankle are wrapped around the horse, and are staying in position in the region of the girth. Her hands have moved forward to allow the horse to stretch out its head and neck over the fence.*

Above: *The rider has stood up in her stirrups before reaching the fence, and is unable to use her legs correctly. At the same time she has dropped the contact through the reins. A horse may well take advantage of this lack of contact to refuse the fence.*

Types of jump

From this point, the only way to improve and develop your jumping is to practice it over as many different types of obstacles in the greatest number of different conditions as possible.

Jumps fall into two basic categories. These are uprights, such as walls, gates, hurdles, narrow hedges, and poles placed in a vertical line on top of one another, and spreads, such as parallel or triple bars and oxers. In addition, ditches and banks should also be included in your practice jumping sessions. In the early stages, concentrate on jumping low jumps well, rather than raising the poles higher and higher and jumping badly. Practice jumps – you can easily construct these yourself with a little imagination – should not be more than about 3ft (90cm) high and many should be smaller. It is far better to increase the width of a fence, so that the horse has to stretch himself over it, than to keep testing his high-jump ability by raising the height – better for him and better for you to get the feel of the jumping movement.

A ground line placed in front of a jump, particularly an upright, will help you and your horse to judge the takeoff point more easily. The takeoff point should be approximately the same distance away from the jump as the height of it. This will vary according to the height of the fence and the speed of the approach; a horse approaching a small jump quite fast, for example, will take off far further in front of it than the height of that obstacle.

If the establishment where you are learning to ride has a jumping lane, or if you are jumping in the confines of a school, a useful exercise is to negotiate a line of low poles with your arms crossed and the reins knotted around the horse's neck. This will show you how much you are relying on the reins to balance you. Pick up the reins quickly at the end of the line; then try riding it again doing something like buttoning your coat or knotting a piece of string. This helps to encourage you to ride by "feel" and instinct.

An even more testing exercise is to quit and cross your stirrups as well as riding with no reins, so that now you can only rely on the balance of your position to keep you sitting correctly. This is a useful and practical exercise, since there are few riders who never lose their stirrup irons at some time in the middle of a jumping course. It is comforting to know that you will not be unhorsed immediately.

COMMON FAULTS

1 It is easy to get left behind over a jump. The horse may surprise you by taking off early, or it may make a bigger jump than you expect it to. The rider copes by slipping the reins, that is, letting them run through her fingers, so that the horse is not restricted at all as it puts in a large jump. This will ensure that she does not catch the horse in the mouth.

2 By looking down and to one side of the horse, the rider is losing her balance, and has allowed her leg to move back so that her leg aid will be less effective. As she is looking down, she will be unable to give instructions on landing.

Jumping doubles

When you have negotiated poles and different types of individual fences, you can move onto doubles. These consist of two fences positioned close together, usually with one nonjumping stride in the middle. As you have to negotiate two fences within a short space of time it is even more important that you maintain the correct position, so that you can communicate effectively and positively with the horse throughout the whole jump.

As you approach a double, aim your horse at the center of the fences and look to the second part. This will stop you looking down or to one side as you go over the first part.

You have to judge the correct amount of impulsion carefully coming into a double. If you ride in with too much impulsion your horse will jump too far in over the first part, and find it difficult to jump over the second part. On the other hand, if you let your horse crawl over the first element you will be leaving it with too much to do to get out of the combination neatly.

In either case, the horse may respond by running out at the second element.

If the combination consists of a spread in and an upright out, it will be more difficult to jump because the spread will encourage the horse to jump big, and it may not be able to collect itself and shorten its stride enough to jump the upright.

When you are building practice fences, bear in mind that if the first element consists of cross-poles, it will help you to come in correctly and set you up well for the second element.

JUMPING A DOUBLE

1 The rider's lower leg is on the girth, and her body is folded forward. Her hands are maintaining a good contact with the horse's mouth.

2 The rider is ready to push the horse on using her lower legs. You must not rest on your hands as you come over the first part of a double, or you will not be ready to correct any steering problems ready for the second part.

3 The rider's legs are on the girth, applying pressure, to give the horse precise instructions on how to approach the second element. There is no hesitation on the horse's part as it prepares for the spread.

4 The cross-poles at the second element help to guide the horse to the center of the fence, and it jumps well lifting its shoulders and tucking its forelegs up neatly. The rider is still looking ahead.

5 The rider has folded forward, her lower leg has stayed in position, and she is looking ahead. Her hands are relaxed but in contact, allowing the horse the freedom to use its head and neck as it jumps.

In pushing onto the second part of the double, the rider is leaning too far forward, and her seat is coming out of the saddle. In this position she will find it difficult to give precise instructions to the horse as they come into the second element.

Coping with a refusal

Generally, horses jump badly or refuse to jump for one of two reasons – either they have been badly schooled or they are being badly ridden. At this point, it is more likely that the latter reason will apply. Always try to analyze what it is that you are doing wrong and work at putting it right. Have you interfered with his stride on the approach, jabbed him in the mouth on takeoff, or shifted back into an upright position too quickly on landing, for instance? Any of these errors might make him reluctant to jump for you. If he refuses or runs out, is it because you were uncertain yourself and did not ride him at the jump as if you really meant him to go over it?

If a horse refuses a jump, ride him in a small circle and come straight into the jump again. Horses that constantly run out at fences can often be discouraged from doing so by building high or elaborate "wings" on either side of the fence.

Always finish with a good jump from both you and the horse – however small it may be. This is the one you and your mount will remember for the next session.

ABOVE: *A horse can sense the way you are feeling, and a bad jump may result from your own lack of confidence. Be patient, and attempt to make the jump again.*

RIDING OUT

Having achieved mastery over the sitting trot, the rising trot, and the canter, together with confidence in your ability to control your horse and use the aids properly, there is no reason why you should not take a break from the formality of the school for a ride or two. Bear in mind, though, that this is going to be completely different from the conditions you have so far encountered. Riding on roads, or even across tracks and fields, is very different from riding around a school under the constant, watchful eye of an instructor – even though a qualified person will accompany you.

Most situations you will meet with can be dealt with by your riding experience coupled with common sense, but this is something that people often seem to lose when on horseback.

BELOW: *Allow time for exercising across open country in your routine. It helps to keep a horse fresh and prevents it from becoming bored.*

It is vitally important to be constantly alert. This does not mean that you cannot relax, but you must be ready for the unexpected. Somebody may suddenly emerge from a concealed driveway or something may flutter in the hedge, taking you by surprise and making your mount jump or shy. Control him gently, talking to him to reassure him, and then turn him, to see whatever has startled him.

Do not underestimate the size of your horse when going through gates, or when skirting parked vehicles. The latter should be given a wide berth, but not so wide that you end up riding in the middle of the street. In the same vein, it is wise to ride around man-hole covers, which can be both slippery and potential hazards for the horse to trip over, but do not take this to extremes by going to the other side of the street. Remember you have to steer and control the horse at all times; he will do what you tell him

to do and, if you do not steer him around a stationary truck, for example, he may well either walk into the back of it or just come to a halt behind it. Try not to get into the habit of expecting him to get you both out of difficulties; you are the one in control.

Coping with falls

Something you are bound to experience sooner or later in your riding career is a fall. It may have already happened during one of your riding lessons, or it may be that the ignominy will occur when you are out for a ride. Ignominy is what it is – nine times out of ten when you fall off a horse, the only thing to be hurt is your pride.

Falls occur in all sorts of ways. They may be a gentle slide to the ground when you have lost your balance in the saddle and have reached the point of no return. They

may be caused by not ducking low enough or in time to avoid an overhanging branch, or they may be a seemingly dramatic toss and tumble as the horse trips, halts unexpectedly, or throws you off his back as part of a display of high spirits.

Whatever the cause or type of fall, try to get up as quickly as possible if you are not hurt, to show your instructor or companions that you are allright. Then, no matter what your personal feelings and wishes are at this moment, get back up into the saddle immediately. This is important for your confidence as well as from the point of view of establishing who is master. If a horse senses you are reluctant to remount, this will bring out the worst in him, and he is likely to behave badly. It may help you to analyze the reasons for the fall; was it that you lost your balance, for instance, or was it the result of some other cause that could equally well be worked at and therefore avoided in the future?

Experts differ on whether or not you should make a conscious effort to hold onto the reins when you fall. Often, you have no choice, as they are wrenched from your hand; equally, you sometimes have no time to think. Within the confines of the school, it is generally better to drop them; the horse is not able to escape, as he might do in the open, and letting go of the reins lessens the risk of the horse trampling you.

If a companion falls off when you are out for a ride, the whole ride should stop and wait for him to pick himself up, regain composure, and remount. Help to catch the horse, if necessary, and hold it still for the fallen rider. If a loose horse decides to turn the escapade into a game and refuses to be caught, never chase after him. This will only heighten his excitement and make him even more determined to evade capture. Instead, try to corner him; when he realizes all escape routes are blocked, he will soon give in.

Even though a hack is obviously less formal than school work, do not let your riding deteriorate as a result. Concentrate on maintaining the correct position at all times and on making the horse go well for you. Practice smooth transitions from a walk to a trot, a trot to a canter, and back to a trot and walk again. Make sure the horse does only what you want him to do at the times you ask him to do it. If he displays whims of his own – perhaps to canter at a spot where he usually does – you must correct him. You are still the boss.

OPENING AND CLOSING A GATE

1 Position your horse parallel to the gate with its head facing the latch, take the reins and whip in one hand and, with the hand nearest the gate, undo the latch.

2 Use your leg nearest the gate to ask the horse to move away from it (the horse will be turning on its forehand).

3 Open the gate far enough for you to pass through, remembering that some horses become upset and try to rush. If you do not give yourself enough room you could get badly knocked against the gatepost, or perhaps be unseated.

4 Once through, position the horse parallel to the gate again so that you can pull the gate shut, and fasten the latch. Always watch where your horse's head is – a horse can quite easily catch its bridle on a gate latch.

Across country

Just as there are rules to observe when riding on streets and roads, there are rules to follow when riding across the countryside. There are also elementary codes of good manners to observe towards any pedestrians you meet.

In the country, always close gates behind you, whether the field they border contains livestock or not. Always pass single file through a gate, making sure you leave sufficient room for your knees – a rap on a gatepost can be extremely painful. It helps if one person on the ride holds the gate open for the others, but remember to return this courtesy by waiting on the other side until the gate has been closed and everyone is ready to proceed.

In many country areas, you are the "guest" of a farmer, in that you are riding over his land. Behave as you would if you were a guest in someone's house, by observing basic good manners. Do not canter or gallop across a sown field; indeed, you should not ride at speed across any field, particularly if the ground is very wet. You could cut it up and damage it considerably. If by any chance you do some damage, knock down a fence, or let some animals out, find out whose land it is and tell them what has happened. It is inexcusable to leave someone else to discover the damage, whatever it is; by the time they do so, it might have worsened.

Show consideration for the others on the ride. The pace should always be adjusted to the least experienced and most nervous rider. Never ask or expect your fellow riders to do things they neither want to, nor are ready to do. It would be like asking you to jump a gate or hedge at this stage. Later, when you have learned to jump, never jump every obstacle in sight; if you do, you will become the farmer's enemy, rather than friend. If, you want to jump, jump only those obstacles that you know you are allowed to and which,

ABOVE: *Hill work is beneficial for improving the horse's strength and endurance whatever area of competition you are involved in. Uphill work is of particular value.*

ABOVE: *When riding uphill at a canter it is important to keep your weight well forward, to allow the horse maximum freedom of movement.*

should you crash through them or knock them down, will cause no serious damage – for example, knocking down the boundary fence of a field containing livestock.

When riding through wooded areas, lean well forward – not back – when passing beneath low, overhanging branches. This may sound obvious, but you would be surprised how many people forget to do it. Similarly, if you encounter a swinging branch, do not release it so that it flies into the horse or rider behind. If you meet pedestrians on a narrow track, slow down and go past them at a walk.

If your horse goes lame for any reason, get off immediately and see if you can determine the cause. The most likely reason is a stone lodged in his hoof. If this is so, remove the stone and he will probably be quite sound again. If you cannot find out the reason and the horse continues to be lame, run the stirrup irons up the leathers, take the reins over his head, and lead him home.

· As you will see, all the points mentioned have been no more than common sense and good manners. Do not lose sight of either of these just because you are sitting on the back of a horse.

ABOVE: *Horses always enjoy the opportunity to relax that riding out provides, whatever the conditions.*

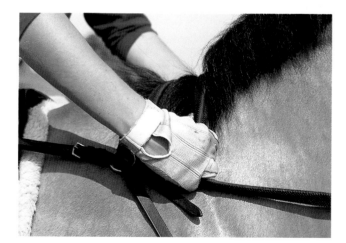

ABOVE: *If you find that your horse becomes rather strong, you will be able to control it better if you bridge the reins by bracing them against the withers. If the horse pulls, it will be pulling against itself, and cannot pull the reins out of your hands.*

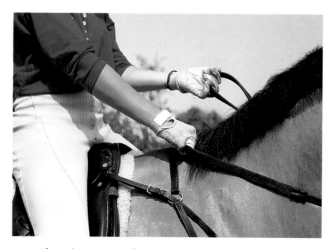

ABOVE: *If your horse is going faster than you are happy with, and your position in the saddle is relatively stable, you can slow it down by anchoring one hand in the horse's mane, and giving a series of short sharp tugs with the other rein. If you pull continuously against a horse, it will only pull harder against you.*

WESTERN RIDING

Contrary to the popular image of Western riding conveyed by the cowboys of the movie screen (galloping furiously across the prairie with legs straight in the stirrups and arms flapping), classical Western riding differs only fractionally from English or European classical riding. The two styles certainly come from the same historical roots, because before the Spaniards landed in Mexico in the 16th century – the first Europeans to reach the continent – the people who lived there had never even seen a horse. Thus, the riding style practiced by the Spaniards – the style followed throughout Europe at the time – must have been the example which the inhabitants of the country learned or copied.

The differences that developed in riding styles came about through practical reasons. The early settlers of the Americas were faced with vast tracts of land which were largely uninhabited, uncultivated, and unfenced, even by natural boundaries. These conditions were entirely new to them, as nothing similar existed in their European homelands and so they were forced to adopt new riding habits. Now, they had to spend hours, if not days, in the saddle, working with their horses to establish farms and ranches. As a result, the saddle was adapted to ensure maximum comfort and greater control. From the native Indians who soon mastered the art of riding on horses that escaped from the early settlements, they learned how to use the lariat to rope stampeding cattle. They soon practiced the technique from horseback and it is now an essential part of Western riding.

Classical Western riding today is still closely linked to the riding of the cowboys. The point to bear in mind, however, is that, to the cowboy, the horse is merely a tool, part of the essential "machinery" of his work. As long as the animal allows the cowboy to do his work with the maximum speed, ease, and efficiency, he is generally not concerned as to whether or not he achieves a classically correct performance.

Schools and differences

There are two recognized schools of Western riding today – the South Western, or Californian, school and the Texan school. In general, the Californian style is more classical, calling for somewhat more refined movements and precision in performance from the horse. Contact with the horse's mouth, although still very light, is more definite than it is in the Texan school. For this reason, the reins are more often very slightly weighted close to the bit – the weighting consisting of no more than a length of braiding of the reins – to make this part fractionally heavier. The Texan school generally demands a little less collection from the horse through the paces, so, to the observer, the horse has a longer "outline" – that is, it is less gathered together.

The most obvious difference between Western and European riding, is that, once horse and rider have been trained correctly, the Western rider holds the reins in one hand only since, quite often, the other hand is needed to hold a lariat. This means that, in order to be able to control the horse, the horse has to be trained to understand the aids and principles associated with neck-reining. In the initial stages, however, it is better for a beginner to hold a rein in each hand. This gives the rider a far greater opportunity to establish a correct and stable position in the saddle. Holding the reins in just one hand

LEFT: *The cinch of a western saddle shown properly tied in a knot.* ABOVE: *the points of the saddle.*

While Western riding may differ only slightly from English or European riding, it was developed much more for reasons of necessity than for esthetic considerations. Above: Western saddles are also unique in having a high horn, deep, wide seat, and high cantle.

tends to pull you out of the saddle and generally plays havoc with what is probably a none-too-secure and well-established position.

The commonly held belief that the Western rider rides with a perfectly straight leg is totally incorrect. To do so would give him no flexibility in his knee or control over his lower leg. Both of these are,

of course, extremely essential in any style of riding.

Clothing and tack

Western riding kit differs markedly from the European equivalent; like the style of riding itself, the kit is the subject of much misconceived criticism. Critics of Western riding claim the clothes are untidy and sloppy – a description which they also apply to the riding style. This is not necessarily so, however; apart from the traditional gear described here, specially designed riding suits are now being worn by more and more Western riders, particulary in the show ring. These are well-cut and tailored suits, designed in the traditional style of Western riding clothes, but with particular attention paid to matching and coordinating colors. Unlike traditional European gear, however, they are often very brightly colored, which has also led to criticism from some traditionalists in the riding fraternity.

The tack worn by horses in Western riding is also different from that used in European classical circles. Once more, it has evolved from equipment developed to suit the conditions of the Western prairies and to give the cowboy every possible assistance and comfort in his work. The bridle is generally "skeleton" in design – that is, it consists of the minimum number of straps and other gadgets. This is partly because leather and metal do not mix well in burning hot conditions, tending to react badly with each other. Bridles traditionally associated with Western riding are either bitless or possess a spade or ordinary Western curb bit: many Western riders, however, prefer to use one of the Western snaffle bits, which is a type of joined mouth curb bit. The

English jointed Pelham is similar, but without a rein for a snaffle bit.

Both horse and rider need to be extremely well-trained if a spade bit is used. This bit has a high port, which moves fractionally in the mouth as the reins are moved. The horse feels this movement on the tongue, and also very slightly on the roof of the mouth. Such bits usually have a curb strap made of leather, rather than the chain one usually used in European riding. This, again, keeps the mixture of leather and metal to a minimum.

There are many different designs of Western saddle, the designs varying to suit the work the cowboy has to do. The type most frequently used for recreational Western riding is known as the Western Pleasure Saddle. This is lighter than most stock working saddles, but still possesses the traditional high horn.

Above: The rider also looks different from the European or English rider because of his riding outfit which would include a wide-brimmed hat for protection from the sun, a brightly colored shirt, blue jeans, leather or suede chaps, and boots with high heels and pointed toes. Again, Western dress is often criticized by European or English riders as being informal and sloppy. However, when one considers the environment and conditions under which most of these riders must perform, the particular Western habit is well-suited to the type of riding.

Mounting

Getting to know the horse on which you are going to learn to ride Western-style is just as important as in any other riding style. Before attempting to mount, therefore, find out the horse's name and pat and make a fuss over him, before leading him to where your lesson is to be. When you are ready to mount, check that the rigging is sufficiently tight to prevent the saddle from slipping.

Mounting Western-style is not very different from mounting European-style. Standing on the horse's near side, hold the reins in your left hand sufficiently tightly to stop the horse from moving forward. Get into the habit of doing this from the start, even though someone will be holding the horse's head during your early lessons. Face obliquely to the horse's hindquarters – that is, not directly toward the tail, as in English equitation, but in such a way that a sideways glance over your left shoulder means you can keep an eye on the horse's head. Turn the stirrup toward you and put your left foot in it. Move your right hand over to the off side of the saddle and rest it against the swell; then spring up off the ground and throw your

right leg over the horse's back and saddle. Remember that the extra height of the Western saddle means you will have to lift your leg slightly higher than normal. Settle into the middle of the saddle and put your right foot in the stirrup.

The advantage of putting your right hand against the off-side swell means first of all that you eliminate any danger of pulling the saddle toward you, as you would if you took hold of the cantle. It also means that you can leave it in this position until you are sitting in the saddle. If you placed the hand further back, you would have to move it forward as you brought your leg across the saddle, which means that, momentarily, you would be balancing in space.

Such precautions as holding the reins sufficiently tight to discourage any forward movement from the horse, and placing your right hand in a way that you do not have to move it while you mount, stem from the early days of Western riding. The worst thing that could happen to a Western rider on the range was to lose his horse. If, for example, it shot forward while he was mounting, so that he lost his balance and let go of the reins, he was often as good as dead. With the horse would go, not only his

ABOVE AND RIGHT: *The rider's position in Western riding is similar to that of European or English. The rider sits evenly in the center of the saddle with the body weight distributed on either side and body weight and thrust directed downward into and out of the heels. As you will be lifting yourself out of the saddle, the stirrup leathers should only be slightly longer than in normal dressage riding. Your head should be held high and straight, body upright, legs lying easily, and hands positioned on the horn or holding the reins.*

transport, but also his canteen of water and his emergency food rations.

MOUNTING

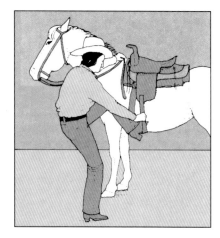

1 Facing obliquely to the horse's hindquarters, put your left foot in the stirrup iron. Hold reins tightly.

2 Hop around to face the horse's side and put your right hand across the seat of the saddle.

3 Jump up off the ground and swing your right leg over the horse's back, settling gently into the saddle.

that, if you stood up in the stirrups, your seat would not lift off the saddle. An exercise which will help you initially to find the right length is to put your feet in the stirrups and let the leathers down until your legs are hanging straight down on either side of the saddle. Remember, however, not to point your toes downward, too. Then, take the leathers up one hole – in most Western saddles, these are positioned about 2in (5cm) apart. If they are set closer together, take the leathers up two holes. Unfortunately, it is impossible to adjust the leathers on a Western saddle when mounted.

The Western seat

As in European equitation, the Western rider sits in the middle and center of the saddle, so that his or her weight is evenly distributed across the horse's back and directly over his center of balance. In just the same way, the rider's head is held high and the weight of the body falls down onto the seat bones. The residue falls down onto the knee and out of the heels; in other words, the rider's weight is all directed downward. The feet should not be braced hard against the stirrups, as this lifts the seat upward out of the saddle.

As mentioned earlier, it is often thought that Western riders ride with a perfectly straight leg. In fact, the stirrup leathers should be about the same length as that used for dressage; that is, slightly longer than for normal, European-style, recreational riding, but not so long

Dismounting

To dismount Western-style, put the reins in your left hand and place them just in front of the saddle. Put your right hand on the saddle horn and take your right foot out of the stirrup. Lean forward and swing your right leg up behind you over the back of the saddle and across the horse's back. Looking toward the front of the horse, step quickly and gently down to the ground. As soon as your right foot touches the ground, take your left foot smoothly out of the stirrup.

DISMOUNTING

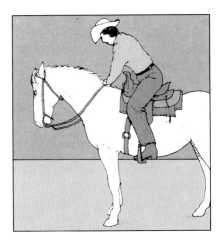

1 Take your right foot out of the stirrup iron, keeping a tight hold of the reins, resting your hand on the horse's neck.

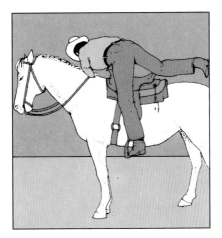

2 With your right hand resting against the right-hand side of the saddle, lean forward and swing your right leg up behind you.

3 Step down to the ground by the horse's side, then take your left foot out of the stirrup iron.

The walk and the jog

As in all types of riding, the most important aspect of Western riding is to get the feel of the horse – to know the feel of sitting correctly in the saddle and to learn to recognize the feel of the horse's legs beneath you. Not only is that difficult to explain but very difficult to teach. The only way to understand even the meaning of this feel is to practice riding continually and extensively. Practise continually the walk, even before you move onto the jog, aiming to achieve smooth turns and perfect circles. Give the aids to turn in exactly the same way as you would normally, feeling gently with the rein in the direction you want the horse to move, with your inside leg pressed against his side close to the cinch and the outside leg applying pressure just behind this. Remember that the pressure you put on the reins should be no more than the slightest squeeze. Your aim is to achieve a smooth, flowing turn, with no jerkiness or violent head reaction from the horse. If you pull on the reins, rather than feel, your mount will inevitably react jerkily and violently.

The Western rider asks his or her horse to move forward into a walk from a halt in exactly the same way as a rider practicing any other style of equitation – that is, by closing or nudging the legs against the horse's side and opening the fingers to allow the horse to move forward. As the horse goes into a walk, the rider must follow the movement by allowing his body to move in time with the rhythm of the pace.

In Western riding the trot is called a jog. The correct jog calls for engagement of the hocks, so the horse is coming from behind with energy and rhythm. Ask for it in the

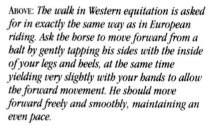

ABOVE: *The walk in Western equitation is asked for in exactly the same way as in European riding. Ask the horse to move forward from a halt by gently tapping his sides with the inside of your legs and heels, at the same time yielding very slightly with your hands to allow the forward movement. He should move forward freely and smoothly, maintaining an even pace.*

usual way, making sure first that your horse is walking out well and is attentive and obedient to your aids. Stay seated in the saddle, as for the sitting trot, relaxing your body so that you are able to follow the horse's movement. Do not brace your body against the saddle – if you do, you will inevitably bounce out of it – but, equally, do not relax enough to become sloppy. Sit up straight and let your loins and waist absorb the movement.

As with the walk, do lots of practice work at the jog, trying to maintain a completely even and steady pace for three or four circuits of the school at a time, and then through smaller 65ft (20m) circles and figure-eights. This is considerably more difficult to achieve than it sounds on paper, but it is extremely good practice. You

LEFT: *The jog is the Western equivalent of a trot. Although the movement is the same as in European styles of riding, the pace tends to be rather more bumpy. The rider should sit deep into the saddle, keeping the waist and loins very supple, in order to allow him to relax with the movement.*

should bear in mind constantly that the Western jog is a collected movement in which the horse's hindquarters are active and tucked well beneath him, giving an alert and active pace.

Traditionally the Western rider always rides at the jog, and not rising like the European rider. However, particularly if you intend to go Western trail riding (pony trekking, Western-style) when you may spend many hours a day in the saddle, it is a good idea to be able to rise to the trot, too. On long rides, this is essential for at least some of the time; sitting in the saddle for very long periods is exceptionally tiring for both horse and rider. You will also find that practicing the rising trot will help you in recognizing feel, as you think more consciously about the legs moving in diagonal pairs beneath you.

Slowing down and stopping

Just as the aids for moving forward, or going from a walk to a jog, are the same in Western riding as they are in other classical styles of equitation, so, too, are the aids for slowing down and stopping. To go from a jog to a walk, close the inside of your lower leg against the horse's side and squeeze on the reins to discourage the forward movement. You should think of "walking" – this advice may sound strange, but, if you think consciously of what you want your horse to do, you are far more likely to transmit your wishes to him clearly. The philosophy is just the same as looking ahead in the direction in which you want to travel, particularly when asking for a turn. Look where you want to go, think about the pace you want from your horse, and the battle is halfway won.

RIGHT: *Again, at a lope, the rider should follow the movement of the horse, by sitting deep in the saddle and relaxing the waist and loins.*

The canter or lope

In Western riding, the canter is called a lope. It is exactly the same pace, although when it is executed correctly, with the horse moving very smoothly, it tends to look a little easier and more relaxed than the European version. Though there is a considerable amount of power coming from the hindquarters, it is a very light pace, so the contact with the reins should be similarly considerate. Such lightness, however, can tempt the horse to fall back into a jog, so the rider must ensure against this by urging him gently forward with the legs all the time.

The aids for the lope are the same aids as in the European canter. If you are moving counterclockwise around the school, you would ask for a lope with the near fore leading; if you are riding round the school on the right rein, or clockwise, your aids should be directed toward making the horse's off-fore lead.

Ride at the lope in the same way as you would at a canter, sitting deep in the saddle with your loins and waist really supple, so that they

can absorb the movement. Once more, you should be sitting bolt upright, your head held high, and looking in the direction you want to go. It is also another pace in which the smoothness and evenness of rhythm can sometimes encourage the unwary rider to become sloppy. When you want to return to a jog, "think jog" and give the aids for a downward transition.

EXERCISES

RIGHT: *The Texan method, the reins are split and not fastened together, they pass through the top of the hand, under the thumb, first, sometimes with the forefinger placed between them. The ends come out of the heel or lower end of the hand. In contrast to the Californian method they are not placed under the free hand, though this still rests on the thigh.*

Many of the exercises described earlier in this section also help the Western rider to establish an independent seat, develop and exercise the correct riding muscles, and co-ordinate body movements to work together and independently, as necessary. Moving the shoulders up and around in a circle is a particularly good exercise, as tension here is a common fault among novice riders. It is also one that leads to problems and mistakes throughout the position. Another useful exercise is to swing your legs backwards and forwards from the knees and then swing the arm backwards and forwards from the shoulder in time with the rhythm of the pace. Try this in walk, jog, and lope. Moving in time with the horse's movement helps the rider to capture and recognize the feel of the pace more easily.

In these early stages – when you are still holding the reins in both hands – practise riding at all paces – turning and circling – without stirrups. Remember your aim is to maintain the position in the saddle at all times, just as if you were riding with your feet in the stirrups.

TOP: *The Californian way of holding the reins in Western riding. The reins pass up through the heel of the hand and out of the top beside the thumb. The spare ends are hobbled together and frequently end in a braided and knotted 'romal'. They lie underneath the free hand on the thigh.*

CHAPTER 4

COMPETITIVE RIDING

DRESSAGE

ABOVE: Blenheim Palace, UK, provides a wonderful backdrop for a dressage competition.

The term "dressage" covers all training done on the flat, and dressage movements and exercises are used for training all horses regardless of which area of the sport they compete in. Its aim is to produce a horse that is strong, supple, well-developed, and obedient to its rider.

You have to teach the horse to move in the different paces, progressing gradually from one to the next. In doing this, you must always take into account the horse's stage of development and training. In the demonstration sequences in this section, most of the movements are shown first by a novice horse that is still learning some of the paces and then by an advanced horse, to illustrate what can be expected at different stages of a horse's development. The advanced horse also demonstrates the self-carriage that all dressage horses must develop. It is essential to build up gradually, as the horse needs to develop the muscles necessary to

perform each action. A horse can be ruined if he is asked to do too much too early.

The rider also has to undergo an intensive training program. You have to work on your position in the saddle. You have to learn to use the leg, seat, and hand aids independently of each other but correctly balanced for the movement you are executing, so that you control the whole horse all of the time. And you need to learn to work with a relaxed but positive mind, as any tenseness or indecision on your part will instantly communicate themselves to the horse. A young horse should be worked for about a half-hour a day, and you should include hill work and grid work in the training routine to get the horse fit and keep its mind fresh. Always remember that your horse is not a machine. Do not drill him, and do not just sit on top and try to dominate him. The horse must want to work with you if you are to get the best out of him.

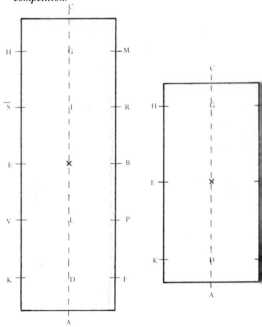

ABOVE: The smaller-sized arena is used for tests up to Medium level, the larger arena for international events. It is useful to practice in a large arena, especially when you are working on advanced movements. The arena is marked with letters which indicate where the movements being performed should begin and end.

LEFT: *If your horse is very lively, you are well-advised to turn it out for a half-hour, or to longe it before beginning a schooling session.*

BELOW: *The aim of dressage training is for horse and rider to learn to work together.*

The walk

In walk, the horse must have good forward movement, be straight, and responsive to your aids. He must be moving forward symmetrically, with the power coming from behind. You must be as still and relaxed as possible and allow, with your hands, to let the horse take a full stride.

There are four types of walk: free, collected, medium, and extended. The free walk is a walk on a long, loose rein. The horse takes long, easy strides and can relax and stretch out his neck.

In collected walk you are looking for increased flexing in the hocks and activity in the joints to give short, elevated steps. The outline should be shortened, and the hind feet should fall just in, or slightly behind, the prints of the front feet. The collected walk is created by using the half-halt repeatedly while maintaining the pace.

The medium walk comes between free and extended. The horse begins to lengthen and extend his stride and the hind legs slightly overtrack the forelegs. He also begins to lengthen his body, although the

outline remains compact and rounded overall. The horse's steps should be free and active, and you should maintain a light contact with the horse's mouth.

In extended walk the horse must just release the outline and become longer and lower. He must cover as much ground as possible with each stride while maintaining regular steps. You ask for the extended walk by increasing the pressure from your seat and closing your legs against the horse. At the same time you allow the horse to stretch his neck forward and take the rein.

ABOVE: *The rider has given the horse the freedom to lower and stretch out its head and neck.*

ABOVE: *By using her legs and seat, the rider asks the horse to lengthen its stride a little. She has a light contact with her hands. The horse's outline remains rounded.*

ABOVE: *In this collected walk (novice), the rider is using her legs to create more activity in the hindquarters, at the same time using a series of half-halts to ask the horse to shorten and heighten its stride.*

ABOVE: *The rider is closing her lower legs to ask the horse to stretch out and lengthen its stride, and allows it to take the rein as it does so. In extending its body, this horse has flattened out too much.*

LEFT: *In this medium walk, this advanced horse has a shorter stride than the younger one. It needs to drop its neck a fraction and lengthen out.*

LEFT: *This advanced horse's hind leg is coming through well from behind and is overtracking the print left by the front foot. Again, it needs to drop a little lower in the neck and lengthen out. Then its hind leg would come through further, lengthening its strides into a good extended walk.*

The trot: working and collected

In trot you should aim for free, active, regular steps, and a general impression of elasticity and suppleness, with the hind legs engaged. It is a difficult pace to do correctly, and young horses, in particular, tend to hurry, causing themselves and their riders to lose their balance.

There are four types of trot: working, collected, medium, and extended. You must be able to follow the movement of extended and collected trot very closely. If your seat is not deep enough, the rhythm of the pace will probably be lost.

The working trot is used for horses that are not ready to learn collection. It demonstrates whether the horse is properly balanced, has good hock action, and is on the bit.

As with all collected paces, collected trot requires plenty of impulsion coming from the hindquarters. You then contain the energy with the half-halt as it comes through from behind, to create springy, elevated paces. The horse's hind legs should be very active, with increased flexion in the hocks, so that they come well under the horse.

The horse's ability in collected trot will depend to a certain extent on his conformation and natural action, as well as his stage of development. Some horses have a very round, natural trot action, and can achieve the required flexion up and hold more easily than others. A horse that has a straight, flat action will have difficulty achieving the same degree of flexion and roundness.

ABOVE: *For this working trot the horse is attentive, it has a good outline, and a nice length of stride.*

ABOVE: *Moving towards collection the horse has good flexion in the hock, and is coming up in front. It is beginning to make its stride shorter and more active, and its outline rounder.*

ABOVE: *The rider has her feet lightly balanced in the stirrups, and her lower legs are wrapped around the horse's body. She is relaxed and independent of her hands, and rises with the horse's movement.*

ABOVE: *Here, however, the rider's body is stiff, and she is leaning slightly backward. Her hands are raised, and the reins are too long, so she has lost the contact.*

ABOVE: *The rider is asking for impulsion and using the half-halt to create a very elevated, collected trot. The horse is flexing well and moving with a rounded action.*

COMMON FAULTS

1   The stride is shorter but the horse has too much weight on its forehand. The rider is slightly forward, and looks as if she is trying to carry it. She needs to sit upright and half-halt to sit the horse down and release and lighten the front.

2   The horse is arguing with the bit. The rider is asking it to start to collect, and is possibly using too much hand, as opposed to using her back and seat properly to ask for the pace. She needs to sit up straight and push the horse forward.

The trot: medium and extended

Medium trot comes between working and extended in length of stride. There should be plenty of impulsion from the hindquarters, and the horse should be taking energetic strides. His neck should be slightly extended. The horse must be overtracking, and the stride must be of equal length in front and behind. Uneven strides show that the horse is not working actively behind.

In extended trot, the horse lengthens his outline to the maximum, to produce long strides. Prepare for the extended trot with the half-halt. Then give strong aids, pushing with the seat and closing the legs firmly against the horse, and release the movement through the reins. Go with the stride as you ask for it. If you do not go with the movement, you will get behind it and end up pulling on the reins.

ABOVE: *The rider is asking for a lengthened stride into a medium trot with her seat and legs, and is keeping a light contact with her hands. The horse is working actively, and the strides in front and behind are of equal length.*

ABOVE: *The rider is fractionally behind the horse's movement. However, as she went with the movement, she is not interfering with the horse's stride.*

ABOVE: *The length of stride is good in this extended trot. The novice horse is still learning this pace, and is going into it too much, with its weight on the forehand. The rider is leaning fractionally forward in an attempt to help it. However, she is allowing the movement to come through. The pace is very active, but a little unbalanced.*

ABOVE: *The rider has asked for the pace by preparing this advanced horse with a half-halt, pushing the horse on strongly, and then releasing the forward movement with her hands, keeping a light contact with the horse's mouth. The horse is responding with a smooth, active, extended stride. You can see how much it has lengthened compared with the medium trot.*

COMMON FAULTS

The horse has its head up, is resisting the bit and going onto the forehand. It is "running" and the rider is having to balance it with her hands. In asking for the pace, she should have used her legs more to engage the horse's hindquarters and push it forward, and then allowed the movement with her hands.

The canter

In canter you are aiming for a general impression of regular steps, lightness, roundness, and acceptance of the bit. The horse should be light in the forehand so that its shoulders are free and mobile, and he should demonstrate good hock action. The horse's back should be round and swinging, and you should be relaxed in the saddle, absorbing the movement. In addition, you must keep the horse going straight, that is with the inner hind foot in line with the inner front foot.

The four canter paces are: working, collected, medium, and extended.

The working canter is used for horses who are not ready to learn collection. It demonstrates a free, balanced pace.

Collected canter should have a clear 3-time beat. The horse should be on the bit, moving forward with its neck raised and arched. He should bring his hindquarters well underneath him, shifting his weight back and lightening his forehand to produce short, springy strides.

In medium canter, again the 3-time movement should be well-marked. The horse should lengthen a little, and his hind leg should come well under his body, producing an active stride.

In extended canter, the hind leg should come well under the horse's body in order to lengthen the stride as much as possible and produce a longer outline. However, apply the pushing aids gently, so that the horse does not go forward in a series of jerks. Extended canter is a difficult pace to maintain because the horse has a natural inclination to rush its strides rather than lengthen them.

It can be difficult to keep a horse straight in canter. If you find this a problem, you should first think about your own position. If you are not sitting square in the saddle you

ABOVE: *The horse is starting collection. Its neck is raised, its hindquarters are rounded and lowered, and it is starting to sit. The rider is slightly against the horse, twisting and pulling back on the inside rein.*

ABOVE: *The rider is in a good position, and the horse is achieving a good stride in this extended canter. It is on the forehand a little because its hindquarters are not quite engaged enough. Although it is lengthening well, it should not drop its head, but must remain in self-carriage as the advanced horse has done.*

will make the horse go crooked. Next, think about the aids you are giving. If you are not balancing them correctly, the horse will not go straight. For example, if your outside leg is too far back, it can make the horse's quarters swing in. If your inside hand is too strong, creating too much bend in the horse's neck, you will lose control of the shoulders and the outside shoulder will fall to the outside.

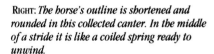

RIGHT: *The horse's outline is shortened and rounded in this collected canter. In the middle of a stride it is like a coiled spring ready to unwind.*

ABOVE: *The stride of this medium canter is a little shorter than the extended canter direct. The rider has simply gone forward with the movement of the horse. She is using her legs to push the horse on and has a good position in the saddle.*

ABOVE: *The horse is making a good long stride in this extended canter, but it is a little flat behind the saddle. It needs to engage the hindquarters more by bringing the inside leg a little further under its body.*

COMMON FAULTS

1 The horse is not using its hindquarters actively enough, with the result that they are not properly rounded. This can be seen behind the saddle, where it is flatter than in the picture above.

2 The horse is on the forehand, resisting the rider. She needs to use her legs to increase the activity in the horse's hindquarters.

Transitions: 1

Transitions between paces and within each pace, for example from collected trot to extended trot, should be executed smoothly and positively, with plenty of impulsion. A rough transition shows poor preparation and lack of balance.

The key to riding a good transition is to have the horse attentive and listening to the aids, and to give a clear, well-prepared request to change pace.

A good way to improve the quality of your transitions is to count the strides as you practice. Ride on a circle, counting the number of strides that you do in each pace, changing the pace on the same count each time. For example, do 10 strides in trot, 10 strides in canter, 10 strides in trot, 10 strides in canter. This will make the horse listen to you. It will also teach you to prepare in advance, because you know that you have to make the transition on the count of 10 each time.

WALK TO TROT

1   The rider is preparing the horse to go forward into trot. She is pushing the horse on with her seat and legs and shortening the reins, and the horse is becoming rounder.

2   The horse moves smoothly into trot. The rider goes with the horse into the new pace, so it is not having to move forward against her hand.

LACK OF PREPARATION

1   The rider is not doing anything to prepare the horse. It is on a long rein and is resisting slightly.

2   As a result, it has taken off into trot and hollowed its outline. The rider has not gone with the movement, so is pulling back on the bit.

TROT TO CANTER

1 The rider feels the inside rein to give slight flexion to the leg she wants leading – in this case the right (inside) leg. Her outside leg goes back behind the girth.

2 The inside leg comes well forward under the horse and the rider moves with the horse as they go smoothly into canter.

ABOVE: *The rider has sat down in the saddle and asked for the transition down to trot, using her legs while squeezing with her hands to check the movement.*

ABOVE: *The rider has failed to prepare for the transition, and the horse is on its forehand, almost falling into the new pace. She should have used the half-halt before asking for the transition.*

ABOVE: *The rider has sat down in the saddle, closed her lower leg against the horse and contained with her hands to ask for the transition from trot to walk. She goes with the horse as it moves into walk.*

ABOVE: *The rider is almost standing in the saddle, and is pulling back on the reins with her hands raised. In response the horse has raised its head, hollowed, and is resisting her.*

Transitions: 2

The canter to walk transition shown here is the most difficult transition to perform. The horse must be listening to the rider and engaging its hindquarters in preparation for the transition. If he is not engaged enough, the horse will hollow, bringing up his head and neck, and will fall back against the rider.

1 The rider has not prepared for the transition, and the horse has its weight on the forehand. Because the horse is a little unbalanced, horse and rider are pulling against each other.

2 The rider is having to use too much hand to bring the horse back into walk. The horse is not submitting to this pressure, but is pulling back to try to escape it.

3 The horse falls into walk, with its weight on the forehand. It is against the movement all the way through.

CANTER TO WALK: ADVANCED HORSE

1 The horse is balanced and collected, with its hind legs coming well underneath it.

2 The rider is pushing down through the seat to tell the horse that the next stride is to be at the walk.

3 The horse moves into walk with its weight well back on its hindquarters.

Turn on the forehand

The turn on the forehand is used in the early part of a horse's training. It increases the horse's suppleness, and teaches him to be obedient to the rider's aids. The movement is also valuable in teaching the rider to control the whole horse.

You ask the horse to fix his shoulders and, using your leg on the girth, you push the horse's hindquarters around. The horse's hind legs step across each other so that he pivots on his forehand.

The turn on the forehand is done from a standstill. Move in a little way from the side of the arena so that the horse has enough space in which to turn.

HOW THE HORSE TURNS

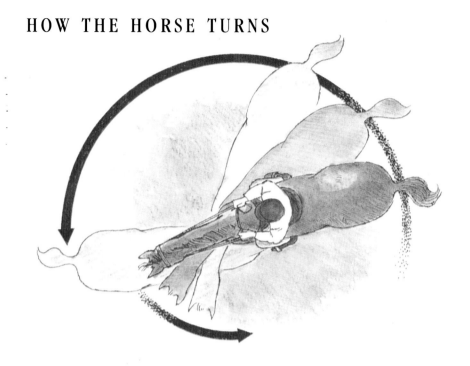

The horse pivots on its forehand through 180 degrees. It is done from the standstill and has little impulsion or foward movement.

TURN ON THE FOREHAND: ADVANCED HORSE

1 Starting from halt, the rider applies her left leg on the girth to ask the horse to move sideways. At the same time she restrains the horse with her hands to keep the forehand in the same place.

2 The horse's hind legs step across each other as it pivots on the forehand. The rider continues to apply the leg aid in rhythm with the movement, while her hands control the horse's shoulders.

3 The horse has turned through 180 degrees while remaining on the same spot.

Counter canter

In counter canter the horse leads with his outside foreleg, and his body is flexed slightly toward the outside of the arena. It should be just as fluid and controlled as the true canter.

The aids for counter canter are the opposite to those used for true canter. The outside leg is used on the girth and asks for strikeoff, while the inside leg is used behind the girth to create impulsion and maintain the movement. The outside rein directs the horse and maintains the bend to the outside, while the inside rein helps to

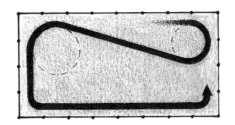

ABOVE: *Ride a small circle and come back to the track in counter canter, then bring the horse back to the walk, or change back to true canter using a simple change. When the horse can do this on both reins, continue the counter canter through the corner of the arena.*

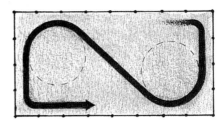

ABOVE: *When the horse can hold counter canter through the end of the school, ride a figure-eight, maintaining counter canter through the whole movement.*

balance the horse and controls his speed and direction.

To begin with, practice counter canter on a shallow loop on the long

side of the arena, maintaining the bend toward the direction of the leading leg. You can then move onto the two exercises illustrated here.

COUNTER CANTER: ADVANCED HORSE

1 The rider has asked for strikeoff in counter canter by using her outside leg on the girth.

2 The rider uses her inside leg slightly behind the girth to maintain the pace, and her outside leg stays on the girth.

3 Her inside hand controls the horse's direction while her outside hand helps to maintain the bend.

4 The counter-canter stride: the outside foreleg leaves the ground as the inside hind leg comes well under the horse.

Shoulder-in

In shoulder-in the horse moves forward and sideways down the track. His forehand is a little inside the track, with his shoulders turned in, while his hindquarters remain on the track. The horse's body is bent around the rider's inside leg, away from the horse's direction of travel, and his shoulders should form an angle of about 30 degrees to the track. You should aim to have the inside hind leg falling in the track left by the outside foreleg, and the horse should be flexed slightly away from the direction he is going. This will give you the correct angle to the track.

Shoulder-in is a good exercise for making the horse more supple. It is particularly good for loosening up a horse's shoulders for jumping. It also teaches the horse to listen to the rider's legs.

The best way to practice shoulder-in is to start it coming off a small circle at one end of the arena. Come around the circle imagining you are going to do another circle. As the horse's shoulders come off the track to start the second circle, resist a little with the inside rein or do a half-halt. Contain the bend with the outside hand so that the horse does not come too far around. Keeping this degree of flexion, push the horse forward down the track with your inside leg on the girth, controlling the hindquarters with your outside leg behind the girth.

THE CORRECT ANGLE

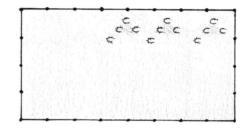

ABOVE: *To achieve the correct angle to the track and to make sure that the horse moves on two tracks, the horse's outside hind leg should fall in the print left by the inside foreleg.*

Shoulder-in is a small movement. Many people tend to think they do not have enough angle to the track, so they overdo it. Experience will teach you the angle correctly.

CORRECT SHOULDER-IN: NOVICE HORSE

1 The rider is using her inside leg to push the horse down the track. Her outside leg is a little behind the girth to keep the horse's hindquarters straight. Her inside hand asks for flexion in the horse's body while her outside hand controls the amount and the horse's direction.

2 The inside hind leg is falling in the print left by the outside foreleg. This young horse is showing the movement well, and has a good angle to the track, but it needs to develop a little more flexion in the body.

STARTING THE MOVEMENT

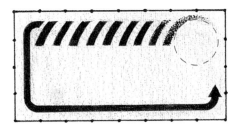

ABOVE: *When starting the movement, imagine that shoulder-in is a segment of a circle, and push the horse down the track, keeping the same degree of flexion.*

COMMON FAULTS

1 The rider is not using her legs and hands to ask the horse to flex.

2 With too much angle to the track, the horse makes too many tracks with its feet. Adjust the angle by controlling the shoulders through the reins.

ABOVE: *In order to perform the more advanced paces well, horses must spend time doing elementary exercises such as shoulder-in when they are young so that they learn to be obedient to the rider's leg.*

The half-pass

In this movement the horse moves forward and sideways down the arena on two tracks. The horse's body should remain parallel to the sides of the arena overall, although it must be flexed toward the way it is going, unlike shoulder-in, when it is flexed away from the direction in which it is going. The straighter hindquarters must neither trail behind nor get ahead of the shoulders. This exercise shows whether the horse is loose in the hips.

To practice half-pass, come around the track and ask the horse to bend in the way you are going by using the inside hand, and controlling the direction of the horse with the outside hand. Your outside leg goes behind the girth to push the horse away, while your inside leg is just on the girth to make sure that the hindquarters do not lead the movement. Look in the direction in which the horse is going.

You must aim for a good, active crossover with the forelegs, with the hind legs stepping out as well, and you must maintain impulsion and rhythm as the horse carries out the movement.

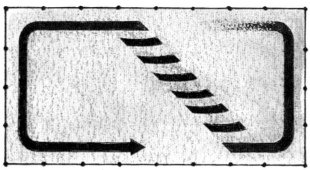

ABOVE: *As you start to come down the long side of the arena, move away from the track toward the center in half-pass.*

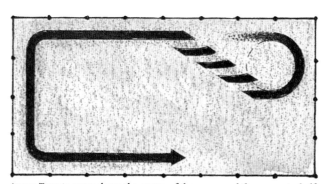

ABOVE: *Turn to come down the center of the arena and then move in half-pass either left or right back to the track. The horse should be straight as you come onto the track again.*

HALF-PASS: NOVICE HORSE

1 The rider is using her outside (left) leg behind the girth to push the horse forward and sideways, while her inside leg controls the hindquarters.

2 The horse is performing this adequately for its stage of training, but is not yet loose enough in the shoulders to perform it well.

3 This horse needs to develop more crossover with the forelegs, and more "expression" in the movement as a whole.

ABOVE: *The horse is using a good through action to make this half-pass, stepping over well with its forelegs. The outside hind is also stepping out well.*

COMMON FAULTS

The horse's hindquarters are moving across the arena fractionally ahead of the shoulders because the rider is not controlling them enough with her inside (right) leg.

ABOVE: *As a horse becomes stronger and more supple, through schooling, it develops self-carriage, or the ability to perform paces and movements with "expression", as demonstrated by this horse and by the advanced horse.*

Leg-yielding

In this movement the horse again travels forward and sideways down the arena. He should have flexion away from the direction in which he is going. As in half-pass, aim to keep the horse parallel overall to the side of the arena. The hindquarters should neither lead nor trail behind the shoulders.

Your outside leg should stay near the girth so that it controls the horse's shoulder. If it goes too far back, it will throw the hindquarters over too much. Feel the outside rein to ask for slight flexion away from the direction the horse is going in. Your other hand controls the degree of flexion.

LEG-YIELDING CORRECTLY

1 The rider is using her outside (left) leg to push the horse sideways, while her inside (right) leg stays near the girth in order to prevent the horse's right shoulder from falling out.

2 At the same time the outside (left) rein is asking the horse for flexion away from the direction in which it is moving. The inside (right) rein is against the horse's neck, helping to support the shoulder and controlling the amount of flexion.

LEG YIELDING: LOOPS

ABOVE: *To ride loops when leg-yielding, ride a shallow loop down the side of the arena leg-yielding towards the center. Go straight ahead for a few strides, then leg-yield in the other direction back to the track.*

LEG YIELDING: CIRCLES

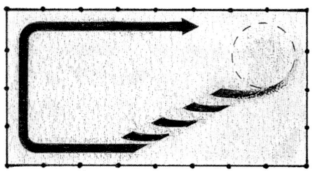

ABOVE: *To ride circles when leg-yielding, ride a small circle, and begin leg-yielding across the arena to the far track as you come off the top of it.*

Walk pirouette

In walk pirouette the horse pivots on its hindquarters. You push the horse as if to move on but bring him around with the reins at the same time. It is done from the walk, and the horse should move smoothly forward into walk without hesitation as he finishes the movement. The hind legs mark time, while the forelegs step across one another. The horse must bend around your inside leg, maintaining flexion in the direction in which he is going, and must pick up his hind feet as it turns.

Prepare the horse with a half-halt and begin the movement as the inside hind stops moving forward. Push the horse on using the outside leg behind the girth, and the inside leg on the girth. At the same time use the rein to ask the horse to step around, while the outside rein controls the degree and speed of the turn.

This exercise should not be done until the horse is working in collected paces and has good impulsion.

A GOOD WALK PIROUETTE

1 The rider is using her seat and legs to push the horse on. She is showing it the way round with the inside rein, while the outside rein comes to the horse's neck to help bring it round.

2 The rider keeps applying the aids in rhythm with the horse's steps. The horse shows good flexion and is looking in the direction it is going.

3 The horse's weight is back on its hindquarters, and its forelegs are crossing over well.

4 The rider is using her leg behind the girth in case the horse's hindquarters are about to swing out.

The halt and rein-back

When a horse becomes more obedient and responsive to the aids for both upward and downward transitions, the halt becomes easier to master.

Although a young horse soon learns to halt and stand still when asked, it is not until he is really accepting the aids that he can perform a true halt from walk or trot. In a correct halt, the weight of the horse should be distributed evenly on all four legs while he remains steadily on the bit and ready for movement forward or backward.

In the rein-back, the horse steps backward with his legs moving in diagonal pairs. He should lift his feet well off the ground and not drag them backward. His weight should be taken on his hind legs without any sinking in the back or lowering of the forehand. He should not resist the hand and should move backward in a straight line. You must be careful not to ask a horse to rein back before he is sufficiently strong in his back and capable of a true halt. Once he has accomplished this, ask him to rein back slightly, lightening your seat and closing the lower leg as if to move forward but, at the same time, restraining the forward movement with your hand. As soon as the horse steps back, reward him with praise and a lighter hand. No more than one or two steps should be asked for in the early stages.

Only when the horse is obedient to the aids should he be asked to step back. If he resists the rein aids, push him up with the lower leg and seat until he realizes that it is simultaneously a restraining but allowing hand. Do not overdo the rein-back, because the horse's back can become tired if it is practiced for too long a period – a few steps each day is quite sufficient. After asking the horse to step back, ask him to walk forward again with light, vibrating leg aids and an allowing hand. The horse must stay on the bit throughout.

Any tendency to run back out of control must be quickly, but quietly, corrected with forward-driving aids and, if necessary, a touch of the whip. Running back is caused by a desire to drop the bit, so the movement should be ridden with slightly more leg aids and a light rein contact throughout. If the horse raises his head and hollows his back when asked for rein-back, then this should be corrected first. In this position, the horse cannot move backward comfortably with your weight on his back, so do not ask for rein-back until the horse has returned to the correct stance.

Care and patience is required in teaching the horse the rein-back. With some fine or thoroughbred horses, you have to take a noticeable proportion of his weight off the horse's back and transfer some of it to his knee and thigh, although without leaning forward. In really difficult cases, it may be advisable to teach the horse the rein-back without your weight. This can be done using long reins, and, once the movement is established, the rider can be reintroduced. This lesson must be learned over a period of time; in no circumstances should it be hurried.

If the horse does not move back in a straight line, this should be corrected with a little more pressure of the leg or rein aid on the side the hindquarters are moving out.

THE AIDS FOR HALT AND REIN-BACK

- To halt, lighten the seat, close lower legs, and push horse forward into a restraining but allowing hand.
- As horse steps back, keep rein aid light and slightly yielding with each step.
- After a few steps, halt, then close legs to ask for forward steps.
- Care must be taken not to hurry the horse forward after rein-back, because this can lead to anticipation and tension in the movement. Ask for a few seconds' halt, every now and then after a rein-back.

The Rein-back

1. Rider lightens the seat.
2. The lower legs are placed behind the girth.
3. Hands are gently restraining, but yielding with the steps.
4. The horse moves backward in two-time, taking diagonal steps.

HALT AND REIN-BACK

1 Here the horse is in a true halt, squarely placed, with its weight distributed over all four legs. It is on the bit and listening for the rider's next command.

2 Now it takes its first steps backward for the rein-back. Note how its legs are moving in diagonal pairs, and the steps are well-raised.

3 Backward movement continues, now on the other diagonal, although the footfall is now not quite in two-time. The foreleg has come to the ground just before the hind leg.

4 Here the horse is better balanced, taking a little more weight on the hind legs, so that the hind foot is now just about touching the ground at the same time as the foreleg.

5 The horse is about to complete the fourth step of the rein-back, after which it will be quietly ridden forward, perhaps after a pause.

Travers

Travers and renvers are movements in which the horse's body is bent in the direction of motion. In travers, the hindquarters are brought in from the track with the horse bent around your inside leg. In renvers, the shoulders are brought in from the track and the horse is bent around your outside leg. Viewed from the front, the angle is greater than that of shoulder-in, so four feet are seen, not three. Travers and renvers are more demanding than shoulder-in and often used with other gymnastic exercises such as shoulder-in and half-pass.

The aids for travers down the long side of the arena are as follows. Before the corner, half-halt, then ride into the corner with the inside seat and leg. As you come out, allow the inside leg to become more passive while the outside leg, placed well behind the girth, asks the horse to move sideways. The inside hand should keep the flexion, with the outside hand regulating the bend and maintaining the balance of the horse. The inside leg should be placed at the girth to give the horse stability, and used more strongly only if the horse's hindquarters come in too much.

Like shoulder-in, travers should sometimes be ridden on the center line, and in fact it is a useful movement when ridden in conjunction with shoulder-in. A good exercise is to ride the horse in shoulder-in, followed by an 8½- or 11-yd (8- or 10-m) circle, a travers, and then a half-circle and half-pass. These movements flow well together and help to establish a positive bend throughout the horse's body, giving him the opportunity to achieve a real balance and rhythm.

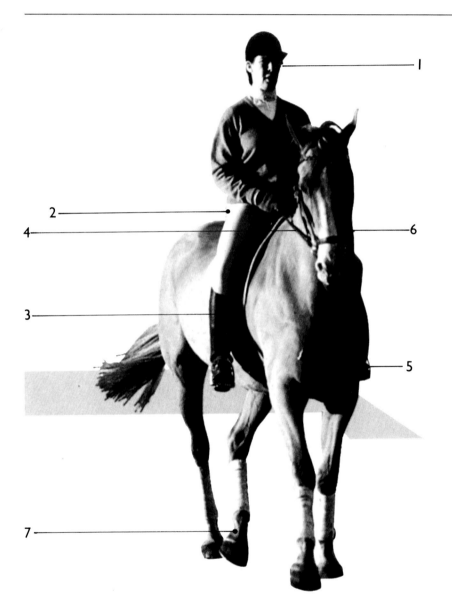

COMMON FAULTS

1 The horse has its hindquarters in from the track.

2 Horse is not bent correctly – its body is straight and only its neck is slightly bent.

3 Rider should place the inside leg on the girth for the horse to bend around.

4 She should sit straight and use much less outside leg – the inside leg has been forgotten.

Travers

1 Rider looks ahead in the direction of the movement.
2 Rider's seat is used firmly but lightly to ride horse forward into the movement.
3 Inside (right) leg is placed on the girth asking for the inside bend, and also riding horse forward into the movement.
4 Right rein asks for inside (right) flexion.

5 Outside (left) leg is placed behind the girth to move the hindquarters into the movement.
6 Left rein controls the balance and regulates the bend.
7 Horse's outside (left) hind leg moves forward and under the horse to cross in front of the right hind leg.

RIDING TRAVERS IN TROT

1 You can see clearly in these pictures that the travers is a four-track movement. Here the outside hind and inside forelegs are on the ground, with the horse well-bent throughout its body.

2 Because the horse's weight is on the inside hind leg and outside foreleg, the bend through the body appears much greater. In fact, this is an illusion brought about by the placement of weight.

3 Here you can see clearly how the outside hind leg is brought forward and under the horse to step in front of the inside hind leg.

EXERCISES FOR TRAVERS

Travers is often performed down the center line, but these movements are useful for training.

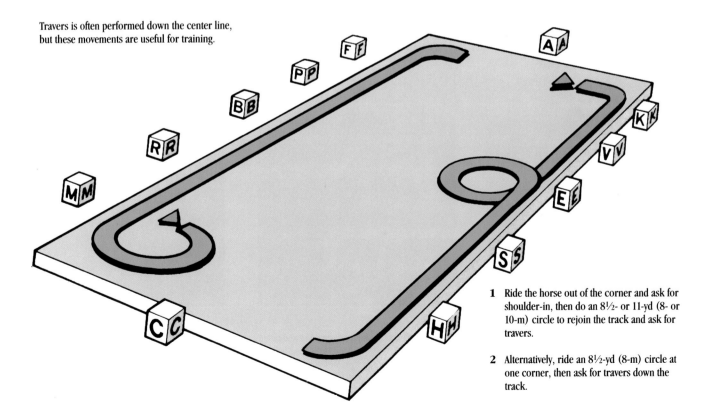

1 Ride the horse out of the corner and ask for shoulder-in, then do an 8½- or 11-yd (8- or 10-m) circle to rejoin the track and ask for travers.

2 Alternatively, ride an 8½-yd (8-m) circle at one corner, then ask for travers down the track.

Renvers

Renvers is, perhaps, a more difficult movement to achieve than travers, because you have to position the shoulders and then ask for the bend when riding it on the track. It is, in fact, easier to ride renvers down the center line, especially if coming out of a 11-yd (10-m) circle, as you hold the hindquarters on the center line, while keeping the shoulders and bend beyond it.

The aids for renvers are as follows. Half-halt at the three-quarter point of the circle, slightly weighting the inside seat bone. Place the inside leg passively on the girth, using the outside leg behind the girth to drive the horse forward and round the inside leg. The hands should regulate the bend in the direction of the movement. Keep the horse forward and on the line with the inside leg.

To ride renvers on the track, start as if riding shoulder-in, then change the bend to the direction of motion by weighting the other seat bone and changing the flexion and bend.

Renvers

1 Rider looks in the direction of the movement.
2 Left leg is on the girth for the horse to bend around.
3 Left hand asks for left flexion.
4 Right leg is placed behind the girth.
5 Right hand controls the shoulders and|the bend, and balances the movement.

RIDING RENVERS IN TROT

1 The horse has just come round the corner of the arena and the rider is asking for left flexion for the renvers. The horse is pictured almost in the moment of suspension in trot.

2 The horse has established the bend and angle of the renvers. The outside (left) hind leg is being placed just in front of the inside (right) hind leg, and the inside (right) foreleg is crossing in front of the outside (left) foreleg.

3 The inside (right) hind leg and outside (left) foreleg are being brought forward and under the horse.

EXERCISES FOR SHOULDER-IN AND RENVERS

Renvers is a useful exercise to perform when a horse is reluctant to finish the half-pass. Ride the half-pass nearly to the track, keeping the bend and the hindquarters pushed over on to the track and take renvers on for several steps. The sequence half-pass, renvers, a reverse half-circle and half-pass again is a useful exercise to establish and reinforce the correct bend throughout all these movements.

1 Begin with shoulder-in (then half-pass) across the arena, leading into renvers to rejoin the track.

2 Or half-pass and renvers: come out of the corner in half-pass, at the track go into renvers, ending with a reverse half-circle. You could go into half-pass again after this, if you wish.

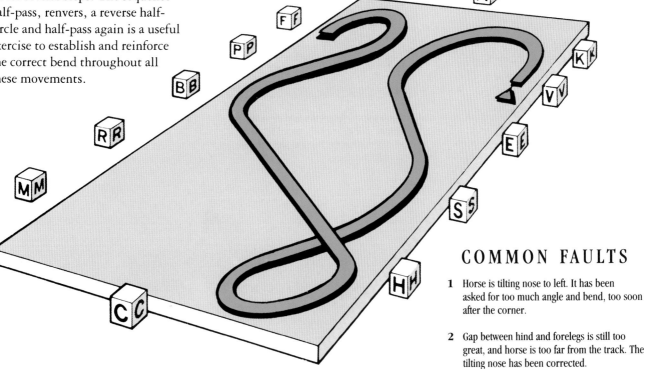

COMMON FAULTS

1 Horse is tilting nose to left. It has been asked for too much angle and bend, too soon after the corner.

2 Gap between hind and forelegs is still too great, and horse is too far from the track. The tilting nose has been corrected.

The flying change

The flying change is a natural pace when the horse jumps from one canter leading leg to the other during the moment of suspension in the canter. Horses with very good, active canters find the flying change very easy to execute. Horses lacking jump in the canter will find it more difficult because the moment of suspension is lacking. You are aiming to achieve lightness and expression in the flying change, and for this you need a canter which is collected, active, straight, and springy.

To execute this movement well, you must be able to feel the exact moment to ask for the change. Make sure your legs are correctly positioned closely round the horse, and, because the change takes place

1 Before making the change, the rider has asked for a half-halt to make the horse pay attention. You can tell by its ears that it is listening for the next instruction.

at the moment of suspension, you must give the horse sufficient warning.

A preparatory half-halt is important to get the horse to listen to the correct aids for a flying change. Move the original outside leg forward just before asking for a change with the new outside leg, which moves back with stronger contact behind the girth. The seat rides the horse on while allowing him to jump, the new inside seat being pushed slightly forward. The legs can then ride the horse forward and straight or give a half-halt if required. The flexion is minimal and is used only to keep the forehand straight. There should be no flexion to the new inside leg, contrary to popular belief. The reins are used only to keep the balance and to give support.

It is important to keep the change straight from the start: swinging the quarters is a bad fault and will affect the horse's later training. A high croup is another fault which must be corrected early in the horse's education: this can turn into a major problem and is an evasion of the aids. With a horse who changes with a high croup, begin by several half-halts, then change with plenty of freedom, riding forward during the actual jump. You may have to

2 The horse has just performed the flying change. The rider has lightened her seat slightly to allow the horse to jump into the air to make the change.

3 The horse has landed on the left leg, and is now in the second stride of the canter. The rider allows it to go forward, while maintaining the left canter position.

EXERCISES FOR FLYING CHANGE

1 In left canter, leave the track and ride a 21-yd (20-m) half-circle: ask for a flying change at point marked X in the diagram. Circle to the right to rejoin track.

2 Circle from the track in right canter, then ask for a flying change at point marked X, and circle to the left. Ask for another change at the lower point marked X, and half-circle again.

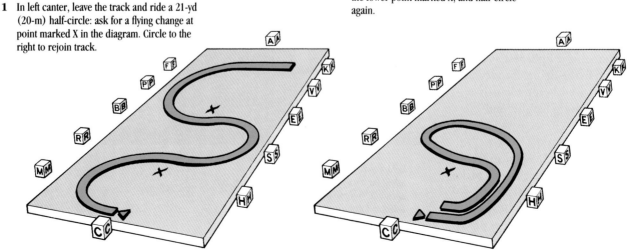

give a half-halt soon after the change because, while correcting the balance of the change itself, he may have lost balance generally. The flying-change aid must be given quietly, but firmly – any throwing of your weight upsets the horse's balance and makes it more difficult for him.

It is easy for a horse to become upset with careless riding of the flying change. You must be sure to keep balance and straightness while releasing the new inside leg sufficiently early, so as not to confuse the horse. If the horse increases his speed into the change, it is likely that the change will be late behind. In this case, more collection is needed. Working on a circle with transitions from counter canter to true canter with simple changes can help eradicate this fault.

Teaching the flying change

When teaching a horse the flying change, using a *manège* can be an advantage. Changing the rein on the short diagonal and asking for the change before arriving at B or E is a good place to start. Alternatively, ride a canter half-pass to the track and, when fully straightened up, the horse will be ready and well-balanced for a flying change. Make sure you ask for the movement well after the half-pass or anticipating could creep in. When the horse fully understands the aids and is happy and settled performing the flying changes while changing the rein and other variations, then practice off the track, changing to counter canter and then back to true canter. At this stage, the counterchange of hand can be practiced in canter but you must always make sure the horse is straight before asking for the change.

Take time to build up collection throughout the whole training of the horse before teaching the canter

pirouette. Transitions in canter are an important part of establishing the required collection. Ride in working canter with many transitions to collected canter. Next, give a few half-halts in collected canter, always checking the horse's straightness by riding in a position right or left. Then ride forward to working canter again. Some steps riding in shoulder-fore in canter help to activate the inside hind leg of the horse; then, from this, the large canter half-pirouette can be asked for. Your chief concern must be the correctness of the steps and the balance of the horse – not the size of the pirouette. Any tendency to throw the shoulders around must be corrected with the outside rein and inside leg, and the pirouette made larger. Once a large half-pirouette can be performed with the horse in balance, you can practice a circle with the horse bent in the direction of the pirouette. Ride travers on the circle, then make a correct circle and ask for a few steps of smaller pirouette, and so on until the horse can accept the collection and balance.

Correct pirouettes are obtained through careful training and not from forced riding. Horses that are short coupled and with a naturally good canter pace find the pirouette

easier than a large, long-backed horse. Any tendency for the horse to raise the forehand and come above the bit in little jumps on the hind legs should be corrected by riding the horse strongly with the legs into the hand and making him come into canter with more energy from the hind legs. Some flexing of the neck before riding the pirouette can help to loosen the horse.

Work should not become too repetitious. Changing the pirouette sometimes to a half-pirouette, a three-quarter pirouette, or even two pirouettes makes the horse listen to the aids.

When the horse is in true collection and balance, the rider can begin to aim for the correct size of pirouette – although even with a highly trained horse the perfect pirouette is achieved only occasionally. Usually, the horse is capable of performing the movement if he is correctly muscled and fit, but it is the rider who needs so much practice in collecting and balancing the horse and in using the aids correctly. Every horse is different, and one may need more inside or outside leg and seat than another. It is this "feel" and sense of timing a good rider develops which is so important when riding advanced dressage.

EXERCISES FOR HALF-PASS AND FLYING CHANGE

Cantering right-handed, turn down the center line, then half-pass out to the track. Canter straight for a few strides, then ask for a flying change at point marked X. This is a very good exercise, because the horse will be well-collected with its hind legs really under it in the half-pass, and with good right flexion.

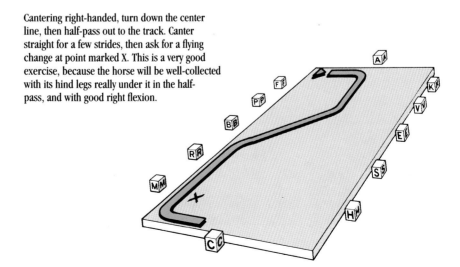

Dressage tests require the horse to be active and free while still displaying all the qualities of power and speed that are its inherent characteristics. It must be light in hand so that the rider can control it with a light contact on the reins and almost invisible aids. It must be supple and obedient, and adjust its paces without resentment. It must remain straight from its head to its tail when moving on a straight line, bending slightly in the direction it is traveling on a curved line. All paces must maintain a regular rhythm with the correct footfall.

In addition, all changes of pace and other movements must take place at the specified markers, your circles must be circles, serpentines must be evenly spaced, and circles and loops must be the specified size.

Competitions are judged by one to three judges, and by five at international level. Each judge has a writer to note down the points scored and comments for each movement. Each movement is marked out of 10 points. In addition, extra points are given for different aspects of your overall performance: for example, general impression and calmness; accuracy of the paces and impulsion; and the position and seat of the rider, and correct application of the aids.

You should aim to arrive early at a competition to give yourself and your horse plenty of time to relax and unwind after the journey. This is not the time for a final practice. Do a few, simple movements to warm up and get the horse listening to you.

After the test study the judges' remarks on your score sheet carefully, and work extra hard at the movements you have been marked down on before the next time.

INTERPRETING YOUR DRESSAGE MARKS

In competition, whatever your level you will be marked on a scale of 0–10. The judges award marks based on the expected performance for the class you have entered.

0	Not Performed.
1	Very bad.
2	Bad.
3	Fairly Bad.
4	Insufficient
5	Sufficient
6	Satisfactory
7	Fairly Good.
8	Good.
9	Very Good.
10	Excellent.

ABOVE: *Horse and rider have a final brush-up and warm-up before going in for their test.*

WESTERN DRESSAGE

Reining patterns

Reining patterns are Western dressage tests. Like their European counterparts, they are executed with great precision, but unlike the former, they are conducted at great speed. The movements called for will vary in order according to the individual test set by the judge.

The tests demand a highly trained horse and a similarly skilled rider. In addition to quick stops, moving into a fast lope from a rein back and flying changes of leg (changing the lead legs at a lope without altering the speed or pace), they also include movements known as pivots, spins and rollbacks.

In a pivot, the horse turns 90° from a halt by pivoting on his hindquarters. The front feet leave the ground at the beginning of the movement and do not touch it again until the pivot is completed; in effect the horse "rocks" around on his hindquarters. A spin consists of a complete 360° turn on the hindquarters. In this the horse is usually brought to a sliding stop from a fast lope. He then executes the spin and immediately moves off on a given lead at the lope. The rollback is also asked for after a horse has been brought to a quick or sliding stop from a fast lope. He must roll around on his hocks, so that he executes a 180° turn with his forelegs coming down into the tracks he has just made. He leaves the turn straightaway on a given lead at a fast lope.

Principles of neck-reining

So far, you have been riding holding a rein in either hand, and you should continue to do this until you have established a really secure position and are thoroughly balanced in the saddle. Balance is of utmost importance, for holding the reins in one hand can unbalance and unhorse a rider very quickly. Make sure, therefore, that you are really competent and confident at all paces – able to maintain an even rhythm of stride and execute smooth turns and transitions – before moving on to ride with the reins in your left hand only.

The first essential to grasp is that you are riding a horse trained to understand the principles of neck-reining. All horses know what it means if pressure is placed on one or other rein; pressure on the left means turn or move to the left, and vice versa for the right. Horses trained in Western equitation go one stage further. They respond to pressure placed on their neck by the reins by moving away from it. Thus, if a rider wants his horse to move to the left, he moves both reins fractionally to the left, so the right rein is pressing against the neck. This is supported with the usual leg aids.

There are two accepted ways of holding the reins in one hand. Try both to see which you prefer and which gets the greatest and smoothest response from your horse. In the first method, the hand is held in the usual way, the wrist slightly flexed and the fingers pointing inward with the thumb on top. The reins are then brought over the top of the hand between the forefingers and thumb. The other end emerges from the bottom of the hand and lies down the left side of the neck. In the other method, the hand is held in the same way but the reins are brought up through the hand from the bottom, emerging between thumb and forefinger. This is a feature of Californian riding.

If the reins used are braided together into a romal (a knotted length of rein made by joining the ends together), this should be placed under the right hand which is laid on the thigh. If the reins are split, the end should be brought up through the hand to fall back over the thumb and the back of the hand, so they are out of the way. They should not dangle down in front or be held under the thigh.

Whichever method of holding the reins you adopt, your right hand should rest evenly and constantly against the thigh. Hold your left hand very slightly in front of the saddle horn – not behind it so it creeps back towards your body. If you take up this position, you will soon have no control over the horse.

Neck-reining in practice

Holding the reins in one hand inevitably tends to encourage various distortions – if not faults – in the position. Therefore practice turning and circling at the walk thoroughly before attempting the jog and lope. Do not overemphasize the technique; although you are asking the horse to move away from pressure on the neck, that pressure should be no more than a whisper. A well-trained horse will respond to just the slightest feel; your hand should never move more than $\frac{1}{4}$– $\frac{1}{2}$in (14mm) as you give the aid. If you move your hand more than this, not only will it affect your balance and position, but it will also probably lead the horse to respond by exaggerating the bend of his neck. He will move his head away from the direction of movement and, if not reined hard to the left, he will be pulled to the right, or vice versa. Think of doing no more than fluttering the reins against the horse's neck, rather than trying to push or force him over.

Supporting leg aids are possibly even more important than usual when neck-reining. The horse must still move forward between your leg and hand, which means that you give the aids with your legs fractionally before the corresponding hand movement. The horse thus moves up to your hand.

When you have begun to master the principles of neck-reining at a walk, move onto turning and circling exercises at the jog. There is no need to exaggerate the aid because the pace is increased; you are still aiming to achieve the steady, even rhythm of the jog throughout all turns and circles.

To get your horse to lope on a given lead, the classic aids are used. Incline the horse's head very slightly in the direction of the lead leg, place your outside leg behind the cinch, and apply pressure with your inside leg. The order of the aids is vital! Although you are holding the reins in one hand, now ask for the bend of the head with the indirect rein – not the neck or direct rein. It will help your horse if you can learn to give these aids as the right shoulder moves forward. By the time the aid has been transmitted to his brain and back to his legs, his next stride will mean that he automatically moves into the lope with the off-fore as the lead leg.

Slowing down and stopping are carried out in the same way as when you are holding a rein in each hand. The action is a gentle closing or squeezing of your fingers on the rein as you ride forward with your legs. However, you must strongly resist the tendency to pull on the rein hand until you find it is almost back into your tummy. This will have no effect on your horse whatsoever.

NECK-REINING ON A TURN

In neck-reining, the horse responds to pressure exerted on his neck by turning away from it. He has been taught to do this in his training. The aids given by the rider should be extremely subtle so the minimum amount of pressure is felt by the horse.

RIGHT: *The rider asks his horse to turn to the right by moving his hand no more than ¼in (1cm) to the right. This is supported by leg aids in which pressure is exerted by the right leg on the girth and the left leg behind the girth.*

LEFT: *To neck-rein to the left, the aids explained on the left are reversed. Once more, the movement of the hand should be barely perceptible to an onlooker.*

COMMON FAULTS

Holding the reins in one hand will affect your position and handling of the horse, especially if you are used to riding with two hands on the reins. Some of the more common mistakes:

Holding the reins unevenly so one is much longer than the other.

Allowing the right hand to come off the thigh and wave in the air, clenching your fist as a sign of tension. The right hand must remain on your thigh at all times.

Holding the left hand too high and too far back.

Moving the rein hand too far to the left and right.

Moving the rein hand forward thereby putting pressure on the horse's neck making him bend back and head acutely.

Leaning into the rein hand, twisting to the right or left. You will eventually rise off the saddle and lose control.

Use of the legs

As well as controlling direction by pressure applied to the neck, you should also be able to make your horse move away from pressure applied with one or other leg. The best way to practice this is through the turn on the forehand, which is a useful movement to learn in any case for everyday recreational riding.

A Western-trained horse executes a turn on the forehand in exactly the same way as any other horse. It also responds to the same aids. Try practicing this exercise to help to achieve a smooth half-turn (through 180°). Place a cavalletti about 1yd (1m) away from the wall or fence of the school in the middle of one long side. Walk the horse on an inside track towards it – on the side nearest to the center of the school. When you reach the far end of the cavalletti, halt with the horse's front legs just clear of the end support. Now prevent forward movement by squeezing gently on the reins, but encourage your mount to move around the end of the cavalletti by bending him round your inside leg. If turning on the forehand, use the leg nearest to the cavalletti in a series of nudges to push your horse's quarters away and so making the turn. When you have completed a 180° turn, walk around the school again to approach the cavalletti from the other direction. Then try the exercise again.

The rein-back

You should also be able to execute a smooth rein-back, which may be required on numerous practical occasions. As in European equitation, forward movement is encouraged by applying pressure with the legs and then restrained by not yielding with the hand. You may find that applying the aids in stages – squeezing very gently on the reins and then relaxing this pressure as the horse takes a step

back – helps to achieve a smooth movement, in which the horse moves back evenly and correctly in two-time. In this instance, the hand aid is "squeeze-yield," "squeeze-yield," as each step is taken.

When the horse has taken three or four steps backward, relax the pressure on the reins and encourage him to move forward. This is to make sure that the impression of moving back does not become fixed in his mind and become a habit. This is important in dressage competitions, when a horse will be severely penalized by the judges if he takes a step back, except when required to do so.

When reining-back, resist any temptation to sit down hard into the saddle in a misguided attempt to induce movement in the horse. This merely makes the horse hollow his back, making it almost impossible for him to step backward correctly. Sit light and think light.

ABOVE: *An experienced rider brings her horse to a quick stop in a competition. The horse stops instantly on command, probably from a lope or gallop, so that his hind legs slide underneath him and his front legs stop dead in their tracks.*

The sliding stop and quick stop

Although a horse should always respond to your aids instantly, even more emphasis than normal is placed on this in Western riding. Again, this originated for practical reasons. If, for example, a rider in mountainous country found himself riding toward the edge of a precipice without realizing it, he would want his horse to stop instantly on a command when realization dawned, not three steps further forward. From this evolved the advanced movement known as a sliding stop, in which a horse stops dead in his tracks – even at a gallop – with his hind legs sliding right underneath him. This puts a tremendous strain on a horse's back and legs, so it should only be asked for very infrequently and then on a prepared surface. If this precaution is not observed, severe grazing of the hind legs can occur. The quick stop is a more common movement. In this, the horse is required to stop as quickly as is physically possible after the aids have been given. He should do so in such a way that he is sufficiently balanced to move off again instantly at any pace.

Transition to the lope

Asking for a lope from a walk, halt or even a rein-back, is a common feature of Western riding. It should be achieved with complete smoothness and precision. To ask a horse to go into a lope from a walk, make sure he is first walking out well and collectedly. Just before you reach a corner, apply the aids for the lope – preferably as his right shoulder moves forward. Think "lope," too. As you approach the corners in this instance, say to yourself "ready" (attract his attention) and "lope" (apply the aids firmly).

The same principle applies when asking for a lope from a halt. Make sure the horse is paying attention and knows you are about to ask him to do something. Shake the reins slightly and nudge with your legs. Then give the aids firmly and definitely, using your legs strongly and yielding with your hands to allow the movement. The common fault is to scoop the reins up toward you, which has the reverse effect from the one you want to achieve. You must keep relaxed, so that your muscles do not tense, and yield with your hands. Such movements are impossible to execute properly if you are in a state of tension.

TURNING ON THE FOREHAND

The turn is a movement in which the horse's hindquarters move around his forequarters, in this case, the horse is describing a complete circle from right to left. Accordingly the rider's right leg is nudging the horse to the left, the pressure from the left leg controlling the speed of the movement. Direction comes from the legs alone, the reins merely restrain forward movement until the turn is completed.

1 The horse is stepping boldly to the left in response to the aids, though there is slight resistance as indicated by the tilted head and flourished tail.

2 The horse has moved through a semicircle. It is nicely settled, with the rider's right leg encouraging the movement.

3 The completed turn, with the horse positioned at the halt ready to move forward.

EVENTING

The cross-country position

A cross-country course is ridden at speed. As the horse gallops around, it stretches out and its center of gravity moves forward. The key to riding a course well is to adopt a forward seat in order to stay balanced with the horse.

You should fold forward from the hip, with your seat out of the saddle, and keep your lower leg securely in position all the time. As you come into a fence, sit lightly in the saddle in order to drive the horse on over it.

It is vital that your legs remain in the correct position, firmly on the girth, all the time, if you are to remain secure in the saddle. If they slide out of position, or flap against the horse's side, you will not be able to maintain a secure position.

Work all the time to balance yourself with the horse, not getting either ahead of, or behind, its movement.

ABOVE: *The rider's weight is forward, and she is well-balanced over the horse. Her lower leg is vertical, and is steady against the horse's side. She is pushing her weight down into her heel. She is maintaining a straight line from her elbow through her hands to the horse's mouth, keeping up a firm contact and good communication.*

ABOVE: *A good, secure position in the saddle will enable you to tackle any type of obstacle with confidence. Here, the rider is leaning back to keep his weight over the horse's center of gravity as they tackle the big drop, but he is still keeping his leg firmly in position on the girth.*

COMMON FAULTS

The rider's leg is too far back, and she is leaning forward to compensate. This brings her ahead of the horse's movements, shifting her weight too far forward and making it difficult for her to apply her lower leg properly. She has lowered her hands, so the line of contact from her elbow to the horse's mouth has been broken.

Gridwork

Gridwork is as valuable in training the event horse as it is in the other areas of competition. It builds the horse's confidence by teaching him to think for himself, and to shorten and lengthen his stride coming to a fence. It is particularly good for practicing combinations, as it teaches the horse to maintain his concentration over a series of fences.

Grids can be used to practice different types of fences and to solve specific problems, as you can vary the jumps and the distances between them. You can bring two fences close together to create a bounce fence. Or you can space the fences apart, with poles on the ground between them, to make an overeager horse concentrate.

Start with one pole and a single fence and build up gradually from there, but do not confront your horse with a mass of poles. It saves a lot of time if you can have a helper to hand, to move fences and put back fallen poles for you.

It is very important that you get the distances in the grid right for your horse. All horses are different, so you need to work out the length of your horse's stride in relation to your own paces, in order to assess distances accurately.

A STRAIGHT FORWARD GRID

1 The grid starts with cross-poles, which guide the horse into the center of the grid, and encourage it to round over the fence.

2 The pole on the ground brings the horse to the second fence correctly. It is looking confident and alert as it takes off.

3 The canter poles between the fences teach an overenthusiastic horse to bring its head down and concentrate, and discourage it from rushing at the next fence.

4 The horse jumps well over the last element. The rider is balanced, with her weight forward, lower legs on the girth, and hands maintaining good contact with the horse's mouth.

SETTING UP A PRACTICE GRID

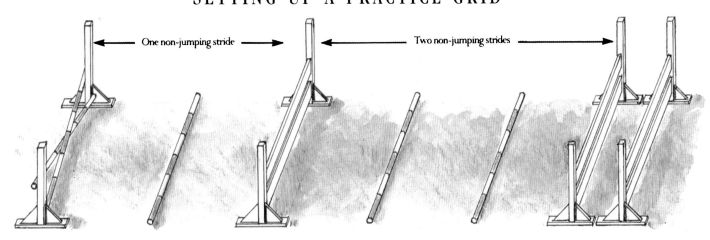

One non-jumping stride ← → Two non-jumping strides ← →

ABOVE: *Practice on this type of grid helps you and the horse to judge the length of stride. The distance between the elements depends on whether you are in trot or canter.*

Approximate distances for setting up a grid		
Approach	*In trot*	*In canter*
Bounce	9-11ft 2.75-3.3m	11-14ft 3.3-4.25m
One non-jumping stride	18-24ft 5.5-7.3mt	24-26ft 7.3-7.9m
Two non-jumping strides	30-32ft 9-9.75m	34-36ft 10.4-11m

RIGHT: *You need to know the length of your horse's stride in relation to your own. You can then set up a grid to suit your own horse by pacing out the distance between each element.*

COMMON FAULTS: RUSHING THE FENCE

1 As the horse rushes at the last fence in the grid, it takes its rider by surprise. She is almost sitting back in the saddle, having been left behind the horse's movement.

2 Because the horse rushed at the fence, it has flattened out over it, raising its head and hollowing its back, and risks knocking it down. The rider is still behind the movement.

Cross-country fences

Cross-country requires more than the ability to jump well. Confidence and courage are needed in horse and rider in order to tackle the wide range of obstacles that are thrown at them. Indecision or fear on your part will communicate themselves to your horse, causing refusals, falls, and halfhearted attempts at jumps.

Jumps are sited in awkward positions, at the top and bottom of steep hills, going uphill or downhill, angled across a slope, or going from sunlight into deep shadow, or an open space into a gloomy group of trees, and vice versa.

It is also not just a matter of jumping uprights and spreads; the jumps can come in many different disguises. Combinations become bounce fences, and may be sited either going uphill or downhill. Corner fences, table fences, steps, jumps into and out of water, even those sometimes on the uphill or downhill, are just some of the many difficult obstacles you will have to tackle. With a little help you can build some of these types of fences at home so that you can practice.

ABOVE: *The corner fence is one of the most difficult on a cross-country course. The rider is encouraging his horse as it struggles to clear the corner.*

ABOVE: *A zigzag should be treated like a spread. It should be approached with plenty of impulsion, and the horse should be brought in close to the base of the fence for takeoff.*

BELOW: *Bruce Davidson on JJ Babu takes a spectacular combination of water and uphill steps. Cross-country obstacles sometimes consist of daunting combinations of different types of fences.*

Bounce fences

On a cross-country course you are quite likely to meet a "bounce" fence. This is a form of combination fence where there is no nonjumping stride between the elements. The horse lands over the first part and immediately takes off for the second.

You can practice riding a bounce either on its own or as part of your grid. As your horse grows in confidence and experience, you could even try a double bounce.

This type of fence places great demands on a horse's athleticism and coordination. As his forelegs touch the ground over the first element, he must bring his hindquarters as far underneath it as possible to take off for the second part.

To tackle a bounce fence successfully, the horse needs to be supple and agile. You need to bring him in on a short, controlled stride, well-balanced and with plenty of impulsion. If the horse comes in long and flat, he will probably land too far over the first element and have trouble with the second. You should sit up slightly between the elements so that you do not push the horse onto its forehand as he lifts himself up over the second element.

With this type of fence in particular, it is very important that your position follows through with the jump of the horse so that you do not get ahead of or behind it. You must keep your weight pushing down into your heels and let your hands go forward with the horse's movement, keeping in communication with the horse all the way.

TAKING A BOUNCE FENCE

1 The rider is sitting down and using her seat, legs, and hands to bring the horse in on a short, bouncy, controlled, well-balanced stride. Its hocks are well underneath it and it has plenty of impulsion.

2 The horse lifts itself up well over the fence, although the rider's position is a little forward of the movement.

3 As the horse lands over the first part, it brings its hocks well underneath it in preparation to spring off over the second part. The rider sits up slightly, applies her lower leg and allows the horse's movement with her hands.

DOWNHILL BOUNCE

1 Horse and rider take off over the first part of a downhill bounce fence. A straight approach and neat, athletic jumping are essential for this type of fence.

2 The rider has her weight back as they land over the first part. The partnership looks confident and well-balanced.

3 The horse has brought its hocks under and taken off close to the second part. They look set to clear it comfortably.

COMMON FAULTS: POOR CONTACT WITH THE REINS

1 Rather than going forward with the horse's movement, the rider has dropped her hands, so that they pull down on the horse's mouth and prevent it from stretching its head and neck forward sufficiently.

2 As a result, the horse has flattened out over the fence, raising its head and hollowing its outline. Its hindquarters are trailing over the fence and might knock it down.

Corner fences

Corner fences can be awkward obstacles when you first meet them in competition. Approached correctly, however, they shouldn't cause too many problems.

Try setting up a practice jump, using a couple of poles and three jump stands. Start with the jump quite low and the angle between the poles quite narrow. As your schooling progresses and you and your horse become more confident, you can open out the angle and increase the height of the jump as necessary.

This type of fence can easily confuse your horse if you are not clear in your approach to it. Cut across the corner and you run the risk of the horse running out (remember that in a competition there won't be a jump stand — possibly just a flag). Approach the jump riding across, but away from, the corner and you still present the horse with problems. He will have to tackle the widest part of the spread, and the front pole will be angling outward toward him, making takeoff difficult.

The best way to tackle this type of fence is straight on but toward the corner, so that the spread is not too great. It is particularly important that you keep the horse going straight and forward for this type of fence, which has a wide spread and can be quite demanding. If the jump becomes an effort, the horse will start to lose confidence.

JUMPING A CORNER FENCE

1   Horse and rider approach towards the centre of the fence. The rider is sitting well down in the saddle and pushing firmly with her legs. She is sticking to her line of approach, and the horse is quite clear about what is being asked of it.

2   At the point of takeoff the horse's weight is well balanced over its hocks. The rider keeps her weight pushing down into her heels as she starts to fold forward.

LEFT: *Approach a corner fence straight ahead but near the corner.*

3 As they go over the jump, the rider's position is good. Her hands are forward but haven't dropped, and she has folded forward from the hip, making it easier for the horse to stretch out over the spread and tuck up its front legs.

4 Horse and rider land well-balanced and ready to ride onto the next fence.

COMMON FAULTS

1 The approach to the fence is on the side away from the corner, and the horse is backing off. The rider needs to push the horse forward more in order to encourage it to lengthen its stride into the fence.

2 Indecisive approach. The horse is heading toward the corner instead of meeting the fence straight ahead and is approaching in a hesitant manner. The rider is not using her whip enough, nor is she sitting down in the saddle. She needs to use her outside leg much more to keep the horse straight, and to maintain firmer contact with the outside rein.

3 Lack of confidence in the approach is communicated, with the result that the horse runs out.

Table fences

These big, solid fences are always found on a cross-country round. Although they appear solid and imposing, they are easier to jump than they look. If you approach them with plenty of impulsion and plenty of control they should ride well.

A table fence should be treated in the same way as a spread. That is, you should come in on a lengthening stride and get close in for takeoff. However, although they have a very solid top line, they usually do not have a proper ground line, and it is easy to come in too close. You need to judge the takeoff point very carefully.

With any type of fence you need to bear in mind that you should always adjust the way you ride to the temperament and ability of the horse you are riding. If you do this, you will be able to help the horse to jump to the best of his ability. If you have an impetuous, overeager horse, sit up a little to control him as you approach a fence, and let the fence come to you. If you are on a more sluggish horse, sit down in the saddle and drive the horse on hard with the legs, in rhythm with the horse's strides, in order to create the necessary speed and impulsion. Fold right down over the fence so that the horse can really stretch out over it.

ABOVE: *Lucinda Green on Shannagh at Badminton, UK. Table fences of all sorts require strong, accurate riding, but are quite straightforward.*

SLUGGISH HORSE

1 This horse lacks natural impulsion, and has to be encouraged over the fence. The rider drives the horse on hard with her legs as they approach takeoff.

2 As they go over the fence she folds down as much as possible to allow the horse to stretch to its maximum, to make up for any lack of impulsion on takeoff.

IMPETUOUS HORSE

1 This horse needs a little encouragement to jump. The rider is asking the horse to take the jump steadily, by remaining slightly upright in the saddle in order to hold the horse's natural exuberance in check.

2 The horse springs out well over the fence. The rider's leg position remains secure and he is folding forward from the hip.

Drop fences

A drop fence, that is, a fence where the ground is lower on the landing side than on the takeoff side, can strike terror in the heart of even experienced cross-country riders.

The secret is to come in very steadily – preferably at the trot – making sure that the horse is balanced and that his weight is well back on his hindquarters. The horse must not be in danger of falling onto his forehand. Bring the horse in as close as possible to the fence, with his hocks well underneath. Keep fairly upright to help the horse lift himself up, and aim for a neat jump over the top. If you take the jump slowly you reduce the risk of the horse jarring himself on landing.

If the horse stands off at a drop fence, he will land a long way out from the fence, where the ground will fall away more steeply. The horse may jar himself as he lands, and an inexperienced rider whose

TAKING A DROP FENCE

balance and position are not quite secure will be thrown forward.

You can easily be thrown out of position over this type of jump if you do not go with the horse's movement all the way. Keep your leg position secure and do not interfere as the horse stretches over the fence. Allow with your hands, letting the reins slip through your

1 This is a cautious horse, and it is coming in steadily. The rider is sitting down in the saddle, and is squeezing with her legs to create an active pace, while keeping good contact, through her hands, with the horse's mouth.

fingers so that the horse can stretch his head and neck as far downward as he needs to, to balance himself on landing.

LACK OF CONTROL

1 The horse is approaching at a trot, but it is "running" at the fence because the rider is not controlling the pace. The rider is too far back in the saddle and is pulling back on the reins. He needs to use his legs to make the horse bring its hocks underneath it, and to contain the horse's movement with his hands, in order to create a more active pace.

2 The rider is pulling on the reins in an attempt to contain the horse's movement. The horse is trying to stretch out against the pull of the rider's hand, and is having difficulty lifting its forehand up over the fence.

2 By pushing on with her legs the rider has made the horse bring its hocks well underneath it, close into the fence, for takeoff. The rider has her weight well-balanced above the horse as it rounds over the fence.

3 Because the horse took off close to the fence it lands reasonably close in, and is not jarred by the landing. The rider is allowing the horse to stretch its head and neck right out in order to balance itself, but at the same time is maintaining good contact with the horse's mouth.

3 The rider has fallen behind the horse's movement, with his weight right back in the saddle. However, he is allowing with the reins as much as he can from this position.

4 Horse and rider have remained in balance, but have landed a long way out from the fence, where the ground is falling away.

Downhill fences

Downhill fences require a similar style of riding to drop fences. However, the approach may be downhill for a while before you reach the jump, and it can be difficult to maintain a good, balanced approach. You should come in on a controlled stride, and always take a downhill fence straight.

It is very important to follow the natural movement of the horse, keeping a more upright position in the saddle so that your weight does not move forward ahead of the point of balance for the horse. Allow with the hands so that the horse can stretch his head and neck down without restriction to keep his balance on landing. You must remain balanced and independent of your hands throughout the jump.

Make sure that you get in close to this type of fence. If the horse stands off, he cannot see where he will land and may be reluctant to jump.

BELOW: *The rider has his weight back when jumping downhill and is giving the horse its head so that it can balance itself on landing.*

1 Horse and rider have made a balanced approach. The rider is driving with his legs as they come into the fence, and the horse has brought its hocks well underneath it. They arrive close in for takeoff.

RIGHT: *The rider has his weight well back as the horse lands on this steeply sloping obstacle.*

2 The rider has stayed slightly upright as they go over the fence and his leg remains in a secure position, keeping him balanced, and allowing him to follow the natural movement of the horse. His hands have a good contact with the horse's mouth.

3 On landing, the rider lets the reins slip through his fingers, so that the horse has complete freeedom to balance itself. His legs have remained in position, and he is completely balanced and secure in the saddle.

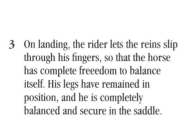

Uphill fences

Uphill fences require a lot of impulsion in the horse, to give him the necessary thrust to bring his hocks underneath and spring up and over the fence.

You and your horse must be well-balanced coming up the hill on the approach, and you must be coming in with a very bouncy stride. You create this by squeezing hard with the legs and controlling the horse's forward movement with your hands.

The key to jumping this type of fence is to keep with the horse all the way up the hill, and to keep your weight forward over the horse as he jumps the fence so that you do not interfere with his natural action.

You can easily get left behind on takeoff, which will throw your weight back over the horse's hindquarters and make it difficult for the horse to spring up off his hocks.

TAKING AN UPHILL FENCE

1 Although the rider has got a little behind the horse's movement as it took off, he is aware of the problem and has extended his arms right forward, in order not to restrict the horse.

2 The rider is still too far back in the saddle but his legs are securely on the horse, and he is well-balanced over it. He is not restricting the horse, which is stretching out well over the fence.

3 The rider has remained secure in the saddle, which has enabled him to recover to a good position quickly on landing.

COMMON FAULTS

The horse is having to twist itself up over the fence because it has come in on an unbalanced stride.

TOP: *Standing off at an uphill fence will make it seem much larger than it is.*

ABOVE: *Uphill fences require bold uphill riding and tremendous impulsion.*

Steps: downhill

Many riders are worried by steps when they first come across them. However, they are really a series of downhill or uphill fences in quick succession, and should be approached in the same way.

Approach downhill steps on a controlled, well-balanced stride. Take them steadily and do not try to do a big jump. Between each step come upright, and put the leg on firmly to tell the horse that you want him to keep going. Keep your hands flexible, following the horse's movement as it stretches over each part.

TAKING DOWNHILL STEPS

1 The rider's contact is good but he is not pushing the horse on quite strongly enough with the legs, so the horse falters at the top of the first step.

2 The horse's hesitation was momentary. The rider is secure and well-balanced in the saddle, staying with the action of the horse. He has come more upright in the saddle and is using the leg firmly to encourage the horse to keep going down the second step.

3 As the horse takes off, the rider has come forward of the movement.

4 The rider has recovered his position in the saddle and lands in balance and with the movement. If you are ahead of the movement, there is always the danger that you will be pitched forward on landing.

ABOVE: *Landing over the first of four down steps horse ana rider look well-balanced and secure.*

Steps: uphill

Steps uphill are jumped in the same way as other uphill fences. You need to have tremendous impulsion, and to drive the horse on with a strong leg so that you do not lose momentum halfway up. Keep your lower leg firmly on the girth and your weight well forward. Once you begin to get behind the movement, you will fall further behind with each jump up.

TAKING UPHILL STEPS

1   As they approach the bottom step, the rider is sitting down in the saddle and is pushing hard with the seat and legs, to create plenty of impulsion.

2   As they land over the first step, the rider has got behind the movement, with his weight a long way back in the saddle. This is restricting the horse's forward movement.

3   The horse takes off strongly up the second step, but the rider's weight is hanging back and his lower leg is starting to creep forward.

4   As they land over the second step, the rider's weight has fallen even further back in the saddle than over the first. However, he is stretching forward with his hands in order to restrict the horse's movement as much as possible.

Water fences

ABOVE: *Bruce Davidson and Dr Peaches land neatly in the water at the Seoul Olympics, 1992.*

Water fences are a common feature of cross-country courses, and the sooner you and your horse get used to jumping in and out of water the better.

Introduce your horse to water gradually. Take him to a nearby stream, having first checked that the stream bed is fairly even and stable, and let him have a good look at the water before you encourage him to go in. Spend some time paddling and walking through it so that the horse feels completely at ease. If you have a second horse with you who is used to water, he can take the lead and encourage the inexperienced one to go in.

Water creates drag as you move through it, slowing the horse down suddenly, and you need to be prepared for this as you jump.

ABOVE: *Horse and rider are well-balanced as they drop down into the water.*

LEFT: *Spray can cause a problem when you are jumping in water. If you can, bring your horse back to a trot so that the spray has a chance to settle.*

If you come in fast and flat to water fences, you will probably be tipped forward as the horse lands and the water checks his forward movement. On the other hand, if you are not pushing the horse on enough, he might stop. Approach them at a steady pace and with plenty of impulsion.

Once you have jumped into the water, sit up and control the horse, bringing him back to a trot if you can. This will give the spray a chance to settle so that the horse can finish the jump.

SHOW JUMPING

ABOVE: *Aim for this shape over a fence. The hocks are deep to the fence, the shoulders are raised high, the neck is arched, the head dropped, and the front legs are folded up well.*

SETTING UP A GRID

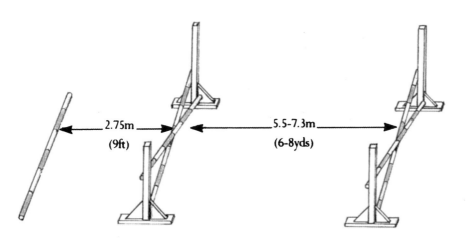

2.75m (9ft)

5.5–7.3m (6–8yds)

A straightforward grid consisting of a placing-pole and two sets of cross-poles. The placing-pole should be positioned about 9ft (2.75m) in front of the first fence, and the cross-poles should be 6–8yd (5.5–7.3m) apart to allow for one stride between them. After practicing over the cross-poles, you can change the second fence to an upright or a spread, because the horse will always arrive at it in the perfect position to jump it.

Gridwork

A grid is a series of practice fences set up at related distances, giving the horse a set number of strides between each fence. Gridwork teaches the horse to set himself up right for a fence and to think for himself, as well as increasing his athletic ability. It strengthens the horse and makes him more supple so that he makes the correct shape over the jump. In particular, it helps the horse to loosen his shoulders and raise them up over the fence, at the same time arching his neck, dropping his head, and folding his forelegs up tight, rounding himself over the fence as he jumps.

You can set up a simple grid consisting of a pole on the ground, the placing-pole, followed by two sets of cross-poles. You should work at the trot until you have developed a good, collected, balanced canter. By trotting over the placing-pole, you ensure that the horse arrives in the correct place to take off for the fence. By arriving consistently in the right place, the horse learns how to set himself up to jump a fence correctly, whereas if you come in at an uncontrolled canter the horse will be too far off one time, and too close in the next. You also do not want to allow the horse to develop the habit of always coming into fences fast and sailing through them on a long, flat stride.

You can shorten or lengthen the distance between the fences to teach the horse to adjust his stride. If the horse gets into the habit of meeting fences correctly, he will start to think for himself rather than depending entirely on your instructions. If he lands a little long he will shorten his stride for the second part and vice versa.

Although gridwork is done mainly for the benefit of the horse, you should practice always maintaining the correct position.

TACKLING A GRID

1 Coming in on a balanced, controlled trot, the horse breaks into a canter over the pole.

2 The rider is in complete control on landing.

3 He increases leg pressure to encourage the horse to lengthen to the second part.

4 The horse lifts its shoulders and rounds well over the second fence.

COMMON FAULTS

The horse has taken off too early. This is making it difficult for it to lift its shoulders and to snap up its forelegs neatly.

Uprights

These jumps can take the form of planks, a gate, a single pole between two jump stands, a series of poles one above the other, or a wall. What they all have in common is that they are built vertically to the ground, and have no spread.

They are the easiest fences to have down because they encourage the horse to flatten out rather than to make a good shape, whereas a spread is built in a way that complements the horse's bascule, encouraging him to round over the fence.

To jump an upright successfully, your approach must be balanced and collected, with plenty of impulsion. If anything, you want to come in on a lengthening stride to teach the horse to get in deep and come back onto his hocks, really using his hindquarters to make a very round shape in the air. Allow the horse's movement with your hands and squeeze the horse up into the air with your legs. If the horse stands off, he will have to stretch to get over the fence, thereby losing the correct, rounded shape. Jump like this regularly, and the horse will get into the habit of jumping flat.

When practicing over verticals, always get in close so that the horse learns to round over them properly. When riding in competition, give the horse a little more room so that he can take the jump in his natural stride but still make a good shape. This will be easier for him, and may also save valuable seconds in a jumpoff. However, don't stand off so far that the horse jumps long and flat.

Verticals are particularly tricky when they come at the end of a course. Horse and rider are eager to finish, and may have been pushing on hard over previous fences, so there is a great temptation to rush them. Be particularly careful to control your approach in this situation so that you jump clear.

TAKE-OFF POSITION

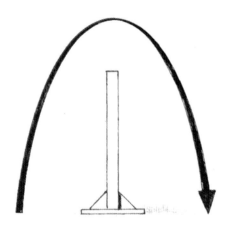

In training, get the horse deep into the fence so that it learns to make a very rounded shape.

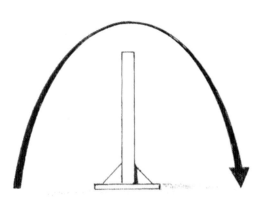

In competition, give it a little more room so that it can take the jump in its natural stride.

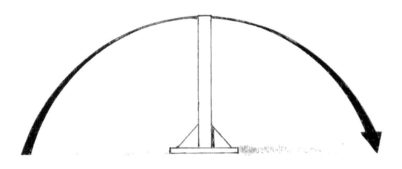

If you takeoff too early, the horse will flatten out over the fence and risk having it down.

ABOVE: *Lisa Jacquin on For the Moment in Paris.*

1   By applying pressure with the legs, the rider
     has the horse coming in on a lengthening, but
     collected, stride, and getting deep into the
     fence. Its hindquarters are positioned well
     underneath it to get the necessary upward
     and forward thrust.

STANDING OFF

1   The horse is taking off too early. It is having
     to reach for the fence, and its front legs are
     dangling.

2   The horse is at its maximum height, but not
     over the top of the upright.

JUMPING AN UPRIGHT

2 The horse has lifted its shoulders well, arched its neck, and dropped its head on takeoff, and folded up its front legs. The rider has folded forward over the jump. His hands have given, but still maintain contact.

3 As the horse prepares to land it maintains its rounded shape. Notice that the rider's leg stays in the same position, applying pressure behind the girth, throughout the jump.

3 The horse is coming down too early. It is dropping its shoulders, raising its head, and hollowing its back in trying to clear the fence.

RIGHT: *Rob Ehrens clearing a typical upright on Olympic Sunrise at Horse of the Year Show.*

Spreads

Spread fences have width as well as height. They include parallel bars, triple bars, oxers, and a wall with rails behind. Fences that slope up away from you are easier to jump than true parallels. For any spread, the horse has to jump wide and high, yet the wider a horse jumps, the less easily he can achieve height.

In order to gain optimum height as well as width over a spread, the horse must get deep into the fence on takeoff and bring his hocks well underneath him. If he stands off, he will clip the top.

You should approach the fence on a deep, lengthening stride. Accurate jumping is essential over this type of fence. You need to be able to see the stride, because the horse must have room to lengthen into the fence. He must not be shortening his stride at the last minute in order to get close to the rail. This ability to see the stride highlights a good ride. It cannot be taught, although it can be developed by practice over poles.

THE CORRECT SHAPE

The width of the horse's jump is measured from the point where its hocks take off to the point where its front feet land. It should make an arc over the fence, the highest point of which must be over the highest part of the fence.

If the horse is supple from flatwork and jumping exercises over grids, he will be able to get his hindquarters right in underneath, bringing his hocks close to the front rail. From this position he springs out over the fence in a good, round shape.

You can get away with standing too far off a parallel only if the fence is not very wide or high – or if the fence is not very wide or high – or if your horse is a brilliant jumper. Do not try to train over large fences. Set up a simple, small parallel spread and practice jumping it perfectly.

LEFT: *The horse has come up strongly off its hocks. It is lifting its shoulders up over the fence and is rounding well. A horse needs to be very supple in the shoulders in order to jump spreads well.*

JUMPING A SPREAD

1 The rider has judged the approach accurately. The horse's hocks are well underneath it and close into the front rail. The rider applies the leg continuously to maintain balance and impulsion.

2 The horse is centrally positioned as it reaches optimum height at the top of the parallel.

3 The horse lands in balance the same distance beyond the fence as it took off in front of it.

COMMON FAULTS: TAKING OFF TOO EARLY

1 The rider has asked the horse to take off too soon, and its forelegs are not folding up neatly.

2 The horse is beginning its descent while it is still over the top of the fence, and is flattening out.

Combination fences: doubles

Combinations are difficult to jump because, when faced with a line of fences, riders tend to panic, which in turn makes the horse panic. As a result they rush at the fence, making it more difficult to come to it correctly. As they go over the first element, they are worrying about the second. The horse raises his head to look at it before landing over the first, causing him to flatten out over the first part and risking bringing it down.

It is up to you to steady your horse and keep his concentration fixed on the element you are jumping until you have landed over it.

Always jump combinations one fence at a time. Do not gallop into them, but ride very firmly. Apply pressure with the leg in order to create impulsion, but do not confuse impulsion with speed. You need to apply more pressure and create a stronger rhythm if the first part is a spread than if it is an upright. Practice jumping the first part really well, to keep the horse concentrating and give him confidence.

Your landing should be controlled and balanced. You should not be leaning on your hands, and they should not move on the reins. Recover your position in the saddle immediately. You will then be able to ask for any adjustments to stride or pace in preparation for the next element.

When practicing, concentrate on achieving a rounded, balanced approach, and teach the horse to take the jumps steadily so that he learns to relax. Do not let him rush on just because there is another fence ahead.

ABOVE: *Neat, controlled jumping, with horse and rider concentrating on the fence at hand. If the elements of a combination are approached in the same way, they should not cause you concern.*

1 The horse has just realized that there is another fence ahead and has raised its head to look at it. As a result it is flattening out as it comes down.

JUMPING A DOUBLE

1 Horse and rider have come in on a controlled stride and the horse is arching well over the first element, an upright.

2 The rider is applying pressure with the leg and the horse is lengthening its stride in order to get in close to the second element, a spread.

COMMON FAULTS: LACK OF CONCENTRATION

2 The horse lands short over the first element, and has a lot of ground to gain if it is to get close enough into the parallel.

3 The rider is completely forward and is really having to push the horse onto reach the second element. They risk standing too far off from it and landing in the middle.

Combination fences – trebles

Exactly the same approach applies to trebles as to doubles. Jump the first element as if it were a single fence. When you land over it, put pressure on with the leg if you need to lengthen to the next element, and contain with the hands if you need to shorten. Whether the second part is a spread or an upright, meet it like an individual fence. Once over it, do whatever is needed in order to meet the third element correctly.

Accurate and balanced riding are more important than ever with a treble fences. You should remember that the faster your approach, the more likely it is that you will have problems because the horse will not be able to set himself up right for the different types of fence in the combination.

As with all practice jumping, keep the fences small and build up the horse's confidence gradually.

COMMON FAULTS: RUNNING OUT

1 The horse is surprised on seeing a third element to jump, and the idea of running out has just occurred to it. It is moving to the right and twisting its body, while the rider is pulling it to the left to try to keep it straight onto the fence.

2 To correct this, the rider sits down in the saddle, and rides with a strong outside leg, so that the horse is in no doubt that it is going to jump the fence.

JUMPING A TREBLE

1 The horse takes off well over the first element, a vertical. The rider's lower leg is squeezing the horse into the air, his hands give with the movement, but they are still in contact with the horse's mouth.

2 Horse and rider are balanced and controlled on landing. The rider sits down in the saddle and applies pressure with the legs to encourage the horse to lengthen into the second element, a spread.

3 The spread has been cleared, and the rider is applying pressure with his legs and containing the movement with his hands in order to meet the third element, another vertical, on a shorter stride.

4 The horse has brought its hindquarters underneath it and rounds well over the vertical.

Spooky fences

Fences with water or ditches under them, fluttering flags, brightly painted poles or planks, or odd colors, can all spook a horse, making him hesitant about jumping them. Very narrow fences, and ones that have little filling-in material, may also worry a horse.

When taking this type of fence, don't make the mistake of galloping at it in the belief that the faster you go, the more likely you are to clear it. If you are approaching at speed, the horse is far more likely to take fright and back off at the last moment, when it sees what he is being asked to jump.

The correct approach is to come in at a slow pace so that the horse can see where you are pointing him and take a good look at what's coming. At the same time, squeeze hard with the legs to create plenty of impulsion and give him confidence.

It is better to come in at a trot and "pop" over the fence, than to come galloping in.

As long as the horse is not frightened by these kinds of fences, he will learn to trust you and respond to your instructions. Construct small versions of some of these fences at home to accustom your horse to taking unusual-looking jumps. For example, you can put a white board on the ground underneath a simple upright to simulate a water tray.

1 On the approach the rider is pushing the horse on in a firm but controlled manner, squeezing with the legs to reassure the horse.

COMMON FAULTS: BACKING OFF

1 The horse's approach is unsure and hesitant. The rider is having to increase the pressure quietly with his legs, to persuade the horse to take the fence.

2 The horse's position as it takes off is crouched and tense, with its hindquarters close to the floor.

JUMPING A WATER TRAY

2 The horse is a little concerned, but is picking itself up well.

3 The horse has been given confidence by the rider's positive approach and it makes the optimum shape with its shoulders over the fence.

ABOVE: *You should prepare your horse for water jumps before meeting them in the ring. When training your horse to jump water, start with a small, inviting jump and increase the width gradually. Do not gallop at this type of fence, but approach it with plenty of impulsion, to encourage the horse to stretch out over it.*

RIGHT: *Elaborate decorations often disguise a relatively easy fence, as here, where giant arrangements of flowers surround a straightforward spread.*

BELOW: *It is not just the height of the wall that makes it spooky. Because it is solid, the horse cannot see what is beyond and may therefore be reluctant to jump it. It requires determined, accurate riding and plenty of impulsion, as well as exceptional jumping ability. This horse has not approached with enough impulsion, and has decided not to attempt it.*

LEFT: *You should prepare your horse for water jumps before meeting them in the ring. When training your horse to jump water, start with a small, inviting jump and increase the width gradually. Do not gallop at this type of fence, but approach it with plenty of impulsion to encourage the horse to stretch out over it.*

RIGHT: *A fence like this should be treated like any other upright fence but approached with very firm riding in order to give the horse confidence.*

Walking a course

Walking the course is an essential part of preparation for the ring, not only to memorize the sequence of jumps, but also to plan how to ride them.

Walk the course as you will ride it, remembering that you should use as much of the ring as possible, except when you are jumping off against the clock. Work out the number of strides you will have coming into awkward fences and those fences that are close together.

In particular, you need to assess the distances in any combinations. To do this accurately, you need to know the length of your own stride, and the length of your horse's stride. You will also need to decide in advance on the rhythm and pace at which you will ride the jump. You can then figure out how many strides you will have between the elements.

If you know that you will have to stretch to reach the second or third

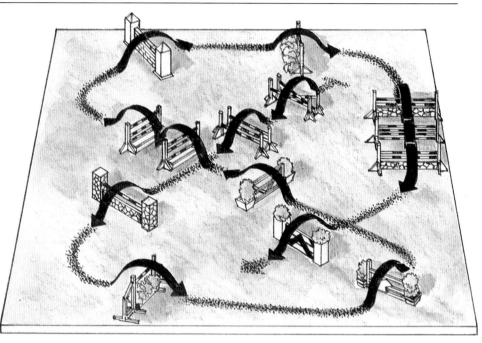

element, you can push the horse on over the first element so that you land over it with the horse going forward, and increase pressure with the leg so that he lengthens out. Alternatively, if the distance will be very tight for your horse, you know that you will have to be ready to collect him back after landing.

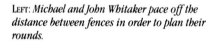

ABOVE: *A show-jumping course will include all types of fences – uprights, spreads, doubles, and trebles – in different forms and combinations. When planning your round, choose a line that brings you in straight to each fence and make use of as much of the ring as possible.*

LEFT: *Michael and John Whitaker pace off the distance between fences in order to plan their rounds.*

OPPOSITE TOP: *John Whitaker (center) checks the fences for stability.*

PLANNING A ROUND

1 Start on the left rein.

2 This vertical needs a collected approach.

3 Be on the left rein here.

4 The upright first element requires a collected approach.

5 Lengthen the stride for the second element.

6 Pace off the distance between the fences to calculate the number of strides.

7 Lead on the left leg here.

8 Collect for this combination.

9 Apply leg to lengthen the stride.

10 This was also the first fence.

11 Pace off the distance. There may be room to shorten the stride.

12 Keep on the right rein.

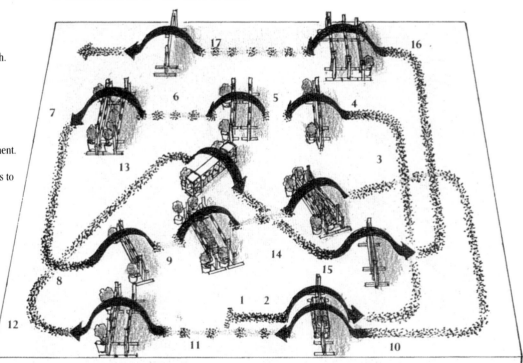

13 Be careful to approach the fence straight ahead after this awkward turn.

14 Make a proper square turn so that the horse has time to see the next fence.

15 The horse may be tired and could flatten out over this vertical. Collect and jump carefully.

16 Approach on a lengthening stride to take off as close as possible.

17 Collect for a steady approach to the last fence.

Riding a course

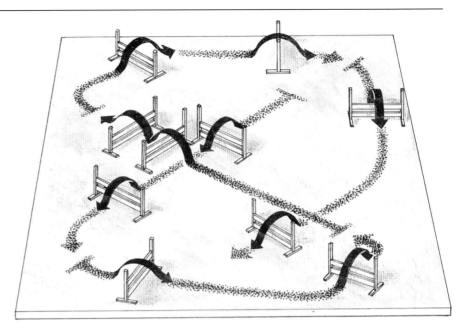

RIGHT: *If you are working a young horse or are new to show-jumping, break the course up into several short sequences of two or three fences, ending in a turn or corner. Treat each sequence as a separate round, aiming for perfect balance and rhythm, going back to trot at the end of each sequence and asking the horse to change leg on the turns. This will teach you to maintain rhythm and consistency throughout the round, and to avoid the frequent mistake of getting gradually faster, flatter, and longer as the round progresses.*

ABOVE: *The bank in the foreground adds an unusual element to this course at Dinard, France.*

It is one thing to jump any amount of single fences, but show-jumping is about completing a course. To do this successfully you need to combine a controlled, rhythmic pace with the ability to execute tight turns and changes of speed and direction smoothly and calmly, but at the same time getting the optimum ability out of your horse over the different types of fence. You should aim for a smooth, controlled round, in which all these elements merge into one fluid, balanced performance.

When practicing at home, aim to be able to canter around ten fences in a controlled way, maintaining the same rhythm throughout the round. A horse only has a certain number of jumps in him, so do not ride him over a course too much at home, and do not practice over large fences – keep them for the ring. Alter the type and sequence of the fences from one session to the next to keep both you and the horse alert.

When setting up a practice course, incorporate several turns and changes of direction to keep the horse balanced and to keep yourself thinking. Experienced horses will automatically put themselves on the correct leg after jumping a fence.

Others, with help from their rider, will perform a flying change. However, don't be afraid to bring your horse back to a trot if you need to, in order to change direction. Then ask for canter again with the correct leg leading.

Think about what you ask the horse to do in the ring. For example, do you ask him to approach fences short, at an angle, or on the turn? Then create these problems over little fences at home so that the horse can learn to cope with them without frightening himself.

ABOVE: *The Olympic Stadium, Seoul, 1992.*

Riding a course (continued)

1   The course begins with an upright. The horse is on the right rein and is being brought in on a bouncy, collected canter, with the aim of maintaining that rhythm throughout the round.

2   After clearing the first fence, horse and rider begin to make a right-handed turn to the next. Ride the turns smoothly to maintain rhythm and to keep the horse balanced.

6   The horse is now leading with the left leg. The rider has asked it to change, in anticipation of a left turn after the next fence, the water tray.

5   The horse jumps out well over the spread. The rider is well-balanced and looking straight in front.

3   Horse and rider make a fairly wide turn to come into the parallel bars. There is no need to cut corners unless you are in a jumpoff, as long as you keep within the allotted time. The horse is leading with its inside (right) foreleg.

4   The turn has enabled them to meet the parallel with a good central approach, well-balanced, and getting in close to the front rail.

Riding a course

7   The rider pushes the horse on with a straight, positive approach and rhythmical, balanced stride that encourage the horse to jump.

8   It clears the water tray without any problems.

9   After the water tray, they turn on the left rein to approach the double. The horse is nicely balanced, maintaining its rhythm, with the left leg leading.

LEADING OFF ON THE CORRECT LEG

You can teach a young horse to lead off from a jump on the correct leg for the way you want to turn. Jump a small fence in a figure-eight, coming in at a slight angle. As you come over the fence, you swing your body weight in the direction you want to go. The horse will soon learn to tell from this which way you are going to turn, and will lead off on the correct leg on landing.

10 They come in toward the double with a good, straight approach and at a steady, collected pace.

Riding a course

11 The horse stands back a bit at the first element, an upright. The rider applies pressure with the lower leg to encourage it over the fence.

MAKING A FLYING CHANGE

This movement will be a great help in the ring. Use your outside leg strongly behind the girth to ask the horse to strike off with that outside hind. By doing this it will be leading with the inside foreleg. At the same time, shift your weight to the inside of the saddle in the direction of the bend. With regular practice the horse comes to recognize these aids and will respond by changing its leading leg.

12 It lands well out over the upright, going forward, and has not been distracted by the second element.

13 The horse has adjusted to meet the parallel correctly, and springs out over it in a good shape.

CHAPTER 5

OTHER EQUESTRIAN SPORTS

LONG-DISTANCE RIDING

The art of long-distance riding is an exacting one; it involves the riding of a carefully trained and conditioned horse over a long distance – "long" meaning anything from about 20 to 25 miles (32 to 40km) up to 100 miles (160km). Such rides may or may not be ridden competitively. As with so many other equestrian sports, its origins can be attributed to the mounted armies of the world, who often had to ride over very long distances in order to reach the place where they were to give battle. In more recent times, many cavalries have conducted "endurance rides" of considerable distances over highly inhospitable terrain to test caliber and stamina.

The greatest following for the sport is in the USA, though its popularity is growing in other countries, such as Britain. In the USA, the sport is known collectively as trail riding. Under this general heading come various categories – pleasure trail, competitive trail, and endurance rides.

Types of rides

Pleasure trail rides need little explanation. They embrace anything from a lone rider taking his mount for a gentle hack across familiar country to a group of riders setting off for a few days with maps, compasses, and overnight equipment to explore new territory on horseback. Route, distance, and speed are up to the riders themselves.

Competitive trail rides, on the other hand, are held across designated, marked trails and those participating in them must complete that course within a maximum and minimum time limit. The rides may be held over one, two, or three days and on the whole they are judged against a standard – that is, horses that complete the course within the set average speed, and which also have no penalty points marked against them at the various veterinary examinations, qualify for the top awards. Others who fall below this average speed, but still have no veterinary penalty points, qualify for a slightly lesser award, and so on. This eliminates any actual racing element, for it means that it is not the fastest time over the course that wins. The most important factors in the judging of competitive trail rides are the overall soundness and condition of the horse at the end of the ride.

Endurance rides are also held over designated, marked trails, but are generally either 50 miles (80km) or 100 miles (160km) long – a distance which must be completed in 12 or 24 hours respectively. Only a minimum time restriction is imposed in endurance rides, so the winner is the person who completes the race in the fastest time with his or her horse in good condition as judged by a panel of veterinarians.

The number of Trail Ride Associations in existence shows how popular the various types of long-distance riding mentioned above are in the in the USA. The most famous US ride is the Tevis Cup – a 100-mile (160-km) one-day ride. It has been held each year since 1955, usually toward the end of July/ beginning of August, the trail running from Tahoe City, across extremely harsh and testing country, to Auburn in California.

Australia and South Africa are currently the two other countries with the greatest number of long-distance riders, although many European countries – France, Germany, and Italy in particular – also have growing interest in the sport. Australia is famous for the 100-mile (160-km) Quilty ride, which is held across the steep, rough terrain of the Blue Mountains in New South Wales. South Africa's principal ride is the National Endurance Ride, which is about 30 miles (48km) longer than the two rides already mentioned. It is held over three days.

In Britain, the best-known long-distance ride is the Golden Horseshoe Ride, which is now organized by the British Horse Society. It is held over two days and covers a distance of 75 miles (120km) across Exmoor, in the southwest. To enter this, riders must either have completed one affiliated 50-mile (80-km) ride during the previous two years, or two of the official qualifying rides. The latter cover a distance of 40 miles (64km) and are held in various places throughout the country during March and April. The Golden Horseshoe is generally held early in May.

Attractions and dangers

To the uninitiated, long-distance riding may not seem to have the instant appeal of the other equestrian sports. It does not appear to embrace the excitements connected with, say, hunting or mounted games; there is no test of jumping ability or dressage skill: no prize money; no exhilarating flat-out gallop, pitting the speed of your horse against others; even the competitive element seems somewhat subdued, in that, in the majority of rides, there is no outright winner. So what is it that appeals to the ever-growing number of riders who participate?

RIGHT: *Riders competing for the Tevis Cup, which takes place in California each year.*

First and most important, those who have gained success in long-distance riding have by no means always been mounted on expensive, well-bred horses. The preparation of the horse for this sport is much more important than the type of horse, it actually is and therein lies its other major attraction. Those who participate regularly in long-distance riding find immense pleasure and satisfaction in the challenge of getting a horse into the proper condition for such a long and tough ride. Preparing a horse for long-distance riding requires extreme dedication, skill, and expertise.

Another often-held misconception about long-distance riding is that it is a sport for the more timid, less competitively motivated riders; those perhaps who find no thrill in hunting, or have not got the "nerve" to gallop around a cross-country event course. However, there is nothing "soft" about long-distance riding – in fact

the reverse is the case. It is an extremely tough sport that demands similar toughness from its participators.

The dangers of long-distance riding can be found equally among its attractions. Because it does not necessarily require an expensive horse or a high degree of skill, as in such fields as dressage or jumping, the unwary could be wooed into participating in long-distance riding without making the proper preparation. Probably because this has proved a problem in the past, strict rules surround participation in official rides, which are imposed to protect both horses and riders. Participants have benefited too from the considerable interest shown in the sport by numbers of veterinarians and, as a result of their studies, new information continually emerges relating to the reactions of a horse under stress.

Horses entering competitive long-distance rides have to undergo veterinary inspections before the

rides, during it (the length of the ride will determine the number of checks), and after it. Temperature, dehydration, pulse, and respiration are the chief factors that the vet uses to judge the horse's condition, so a rider must be familiar with these aspects. In addition, the veterinarian will be looking for any signs of lameness, injuries – particularly those caused by badly fitting saddlery or the horse brushing or over-reaching – and at the condition of the feet and shoes. Horses entered in long-distance rides are allowed no medication or drugs of any type (including the controversial painkiller butazolodine).

Riders therefore must be well-acquainted with the long list of rules before entering any ride. They cover such topics as qualification for entry, saddlery, and tack inspections, ancillary equipment (e.g. some ride organizers ban the use of any leg bandages or boots), shoeing, age, and height of horse, riders' dress, helpers, speed restrictions and allowances, and so on. Currently, these rules vary from ride to ride (according to the organizing body) and from country to country. However, there are plans for some degree of standardization, as the *Federation Equestre Internationale* (International Equestrian Federation) have recently taken an interest in the sport and have now drawn up a list of guidelines.

RIGHT: *A steep climb up Cougar Rock during the Tevis Cup. The rider leans well forward in the saddle in order to take as much weight as possible off the horse's quarters. She is wearing leather or suede chaps over her jeans and a broad-brimmed hat as protection from the sun.*

LEFT: *The long-distance horse must have the stamina and the strength to combat the harshest of terraines. This trail in Arizona passes through rough and arid land, and the horses must find the easiest route.*

The long-distance horse

Most horses should be able to tackle the shorter competitive rides, providing they are sound in limb and wind and that they have undergone a suitable training program to ensure that they are sufficiently fit. If, however, you are looking for a horse specifically for long-distance riding, you should bear the following points firmly in mind.

Good conformation is more important than pretty looks, and probably the most important aspect are the legs and feet. Look at the feet first; they should be a good and uniform shape. The horn should show no sign of splits or cracks: the sole should be slightly concave and tough so that it does not bruise easily. The heels should be fairly wide apart and the frog well-informed, tough, and very resilient. The frog's continued ability to act as a shock absorber over the varied terrain of a long ride is of paramount importance, so examine it carefully.

Experienced long-distance riders will take particular note of the angle of the pastern and the shoulders, looking for a good "slope" in both cases. Look too, for a longish forearm and a short cannon bone with a large, flat knee. The hocks should be well-informed, fairly big, and "well let-down" – meaning, again, that the cannon bones are short.

Viewed from the front, check that the chest is a good width. If it is too narrow, the legs will be too close together, increasing the likelihood of the horse brushing one leg against another; if it is too wide, the legs are likely to swing out awkwardly in

ABOVE: *Competitors will sometimes ride together, at least for some of the way. Quite often there is no prepared track and the riders generally have to find their own route.*

ABOVE: *Narrow chest*

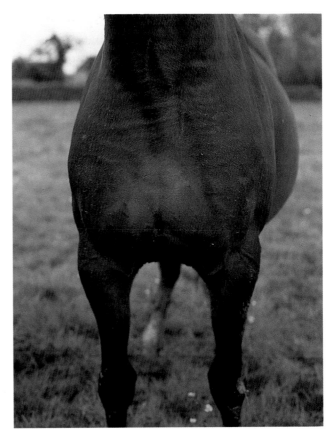

ABOVE: *Good chest*

movement. Avoid toes or hocks that turn inward or outward.

The ideal horse should be short coupled – that is, fairly short in the back. As always, the quarters are extremely important, for they house the animal's driving power. They should be strong, well-rounded, and muscular. There should be a good depth of body from the back down to the girth, to give lots of room for the heart and lungs. The area over the ribs should be well-rounded rather than flat.

The withers and saddle-carrying area are extremely important. The withers must be well-defined; if poorly defined, they will not hold the saddle in the correct position. The back should dip slightly behind the withers so the saddle fits snugly and, again, the ribs should be well-sprung (rounded) so the saddle will not slip backward. The elbows should be well forward of the girth area; if they are tucked into the animal's sides, this could lead to the girth pinching and causing sores. Check all around the saddle area – under the chest as well – to make sure there are no blemishes or evidence of old sores. Any tiny little spot will magnify a hundredfold when work begins in earnest. Excessively thin-skinned horses should be avoided, as wearing tack for long periods could cause them problems.

The head, neck, and forehand should give the appearance of lightness. The neck should not be short and heavily muscled, as this tends to make the horse heavy on its forehand and also prone to pulling. A longer neck gives better balancing ability; but it must not be too long proportionally, as, again, this can make a horse heavy on the forehand. The head should be the right size for the rest of the body and the area under the throatlatch should be wide, so that the passage of air to and from the lungs is in no way restricted. Similarly, a horse with large nostrils has a respiratory advantage. Avoid a horse with heavy, "rounded" bones, giving him an overall coarse appearance. Horses that are very wide or appear to have heavy muscles do not move as easily as those that are lighter in build.

How a horse moves is extremely important. Watch him walking toward you and away from you, and reject any horse that shows even the slightest tendency to brush against any part of one leg with the other, or to overreach. In long-distance riding, this tendency will become more apparent as the horse gets tired, so that toward the end of a ride, a serious, open sore will have developed.

A long-striding horse that moves with relaxed, well-balanced, comfortable paces is the ideal. The more ground he covers with each step he takes, the less quickly he will tire. Look particularly at the walk: does he walk easily and freely or is he constantly wanting to break into a jog? Reject him if he does and reject him, too, if you have to keep kicking him to keep him up to the mark. As much of a long-distance ride will be taken at a trot, make sure this pace is easy and relaxed and comfortable for you to ride. Ideally, the horse should feel equally comfortable when ridden on either diagonal, although this could be something that you will have to achieve with patient training. Try, if possible, to assess the horse's natural cruising speed at a walk and a trot. For most long-distance rides you want a horse that can travel 8 miles (13km) per hour, without coming under stress. Time him for some distance at his natural walk and trot (without pushing him), and calculate an average from this, allowing for some cantering, too.

Many horses experience difficulty in going downhill, taking labored, braking steps, rather than stepping out freely. Obviously this will be a great disadvantage to the long-distance horse, as such a movement will become more overt the more tired he gets. When you are trying a horse, therefore, jog him down a slope to see if it is an effort for him. If each step of the hind legs is a braking step – that is, taken in such a way as to prevent him going faster – he will quickly tire the muscles in the hindquarters and this can result in extreme stiffness or "tying up" of the croup muscles. A veterinarian will eliminate a horse from a ride if it shows such symptoms.

By the same token, you want a horse that is sure-footed and shows no signs of stumbling under any circumstances. A long-distance riding horse should be able to pick his way easily across all types of rough ground. He should have a natural instinct for safety, which, in this instance, manifests itself by not taking risks in tricky conditions.

Temperament and attitude are vital factors. You want an animal that appears always willing to take one step more. On the other hand, he should have a relaxed, calm temperament, as ability to relax, particularly in a strange environment, is extremely important. Make sure the horse you choose is a good traveler, too.

An ability to get on with other horses, to ride along with them at the usual steady pace, and not to get upset when they surge or fall behind is also essential. A long-distance horse must be able to travel alone or in company with equal equanimity. Equally, he must be relaxed with strangers, accepting, say, the veterinarian at the various examinations without any fuss.

As always, have any horse you are considering buying thoroughly examined by a veterinarian; try to find one that understands the stresses imposed on long-distance riding horses. He should take blood tests as well as check the horse's teeth, soundness, and the condition of the heart and lungs.

CONFORMATION FAULTS

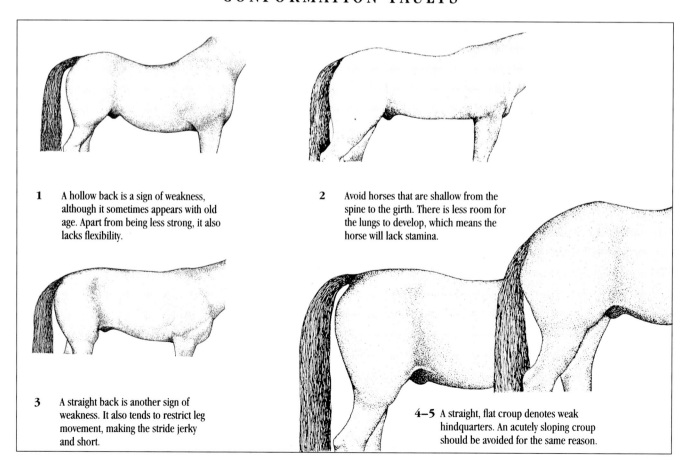

1   A hollow back is a sign of weakness, although it sometimes appears with old age. Apart from being less strong, it also lacks flexibility.

2   Avoid horses that are shallow from the spine to the girth. There is less room for the lungs to develop, which means the horse will lack stamina.

3   A straight back is another sign of weakness. It also tends to restrict leg movement, making the stride jerky and short.

4–5 A straight, flat croup denotes weak hindquarters. An acutely sloping croup should be avoided for the same reason.

HUNTING

Hunting is the oldest of all equestrian sports and, although uncompetitive in itself, it is the forefather of many modern, competitive ones. Ever since man first realized he could cover more ground, faster, and with the extra advantage of the height afforded by sitting astride a horse, he has hunted animals on horseback. Wild boar, hares, foxes, wolves, bears, stags, buffaloes, lions, and even elephants all, at some time or other, have been hunted in such a manner.

In many instances, such hunting has been due to necessity – to provide food or to protect family, livestock, and crops; in many more it has been purely put down to the excitement and thrill of the chase. Robert Surtees, the British 19th-century writer, creator of the widely famous, almost legendary character, Jorrocks, summed up one of the greatest attractions of hunting with the words that it was "the image of war without its guilt and only five-and-twenty per cent of its danger."

Hunting, as it is known today, originated in Britain. The pursuit of game for sport, using packs of hounds to find, chase, and kill the quarry, was certainly well-established in Britain by the time of the Norman Conquest. Although it was thought that the "sportsmen" of the day bitterly resented the preservation of game and the Forest Laws which the Normans were to introduce, these, in fact, brought some order and organization into the hunting field. Hunting began to become an art and a science, as well as a sport.

The principal quarry in those early days was the stag; the fox was at this time looked down upon as not being worthy prey for the mounted followers of the hunt. It was not until the 18th century, when enclosures began to change the nature of the land – and thus also

the business of riding across country after hounds – that fox-hunting began to gain a strong hold. It was soon realized that the crafty nature of the fox presented hounds and huntsmen with a challenge that provided a fascinating day's sport for all; equally good as, if not better than, that given by the stag or hare of the early days.

It was around this time, too, that hunting began to take its present form, with the founding of more organized "packs" of hounds, each of them having clearly defined boundaries across the country, within which they could pursue the sport. Many of the famous British hunts of today – such as those of the counties in the Midlands – were formed at this time.

Europe, the USA, and Australia

Hunting from horseback has been popular in Europe for a long time, too – one of the best examples being stag-hunting in France. Like hunting in Britain, it dates from hundreds of years ago and, although hunting there has followed a somewhat checkered path throughout its history, there are now over 100 well-established packs of hounds in France.

Hunting also has a wide following in the USA and Canada. Its origins in these countries can be traced to Robert Brooke, the son of a British MP, who sailed across the Atlantic in the middle of the 17th century, taking not only his family and servants, but also a pack of foxhounds. Hunting gradually flourished, with the native gray fox being the early quarry, at least until the red fox – which is said to result in a better hunting experience – was imported from Britain at the end of the 19th century.

ABOVE: *Hunting of all sorts has always been a popular subject for artists. Here, a nobleman returns from a successful day's stag-hunting.*

The first packs of hounds were all owned privately; farmers often kept a few hounds for their personal amusement. Toward the end of the 18th century, however, hunting clubs began to be formed, and these mainly corresponded to the British hunting packs. The British influence, in fact, has been the chief one in US and Canadian hunting developments; throughout their history, hunts and hunting in the USA and Canada have been based on the customs, traditions, and techniques found in Britain, although the vastly different types of country and lifestyles have led to various changes over the years.

Australian hunting is based on British precepts, too. However, the quarry there is not confined to the fox; in some parts of Australia, kangaroos are the major quarry.

Hunting today

When used in the context of hounds chasing a quarry, followed by mounted followers, the term "hunting" refers to fox-hunting, stag-hunting, hare-hunting (harriers), drag-hunting, and hunting with bloodhounds. Of these, fox-hunting is by far the most popular in Britain, Ireland, and the USA. There are more than 200 packs of foxhounds in Britain, for instance, while all other types of hunting mentioned have only a handful of packs around Britain (three packs of staghounds, packs of harriers, packs of draghounds, and three packs of bloodhounds, one of which is presently based in the Isle of Man).

Although fox-hunting is also the most popular type of hunting in the USA, with more than 100 packs of hounds, drag-hunting also has a considerable following. It accounts for 15 to 20 percent of all hunting activity. In western states, where foxes are very rarely found, the coyote is hunted. These wild dogs usually run in pairs and are reputed to give a tremendous day's sport. "Blank" days, when hounds fail to find a quarry all day, are rare, but so, too, is a kill, for a coyote will normally outrun the pack. The only time when a kill takes place is said to be when the coyote is "sick or full of chickens"!

Stag-hunting is still the mainstay of hunting with hounds and horses in France as previously mentioned, but the French also hunt wild boar and hares from horseback. Foxes are usually shot. There are a few packs of foxhounds in both Italy and Spain, and a single pack of bloodhounds in Germany.

Procedure and seasons

ABOVE: *Snowy conditions do not prevent hounds from hunting, but they occasionally mean that followers have to take to their feet.*

All hunts have the same hierarchy – that is a Master (a position often shared by two or more people), a huntsman (who may also be the Master) and a couple of Whippers-in (or huntsman's assistants). However, the procedure for each type of hunting and the techniques differ greatly.

At a fox hunt, the hounds are taken from the meet, put into a "covert" (a wood), or led across moors, pasture, or whatever the country comprises, and encouraged by the huntsman to scent a fox. Once they have found a line, the chase is on and they will follow where the fox leads until they either catch it and kill it, lose the line (because of poor scenting conditions, or because the fox is particularly cunning), or run it to ground. If they run it to ground, it is usual to dig the fox out (unless the earth is thought to be very large), after which it must be shot before being given to the hounds. How long it takes for hounds to catch a fox varies every time they find one; the fox may lead them across country for most of the day and still remain free at the end, or it may be caught almost at once. The huntsman will then try to put hounds onto another line.

By and large, fox-hunting comes under the auspices and jurisdiction of the Master of Foxhounds Association (there is also an affiliated US association). Most packs of foxhounds want to be recognized and registered with this organization. Without it, they are not authorized to hold a point-to-point, which is generally a major source of funds, nor may they purchase new hounds from recognized packs. This means that they are not likely to have really top-quality hounds. The Master of Foxhounds Association lays down certain rules connected with hunting and if a pack is found not to be adhering to them, it can face a severe reprimand or expulsion from registration. The first, and most important, of these rules is that the fox must be hunted in its "wild and natural state" – no catching of foxes, to let loose when hounds arrive – in effect, no hunt at all!

Hunting seasons naturally differ from country to country. The season for fox-hunting in the UK starts on November 1 (after that date all foxes are deemed to be "foxes" rather than cubs) and ends usually sometime toward the end of March or early April. Stag-hunting enjoys a much longer season than fox-hunting. It starts at the beginning of August, with the hunting of the "autumn" (fall) stags (these are the biggest); this continues until toward the end of October, at which time there is a rest period of about 10 days. From the beginning of November until the end of February female deer (doe) are hunted – this is the only period when it is legal to pursue a doe. After this, there is another 10-day break and then "spring" stag-hunting carries through until the end of April.

The hunter

Any horse that carries its rider across country, following hunting hounds, may be described as a hunter. There is no specific breed known as a hunter in the way that there is a Thoroughbred, a Quarter Horse, or a Percheron, for example, and a glance around the members of any hunting field will show you that horses of all sizes and shapes can be hunted quite satisfactorily. Nevertheless, thhe fact that certain qualities and characteristics are known to be desirable in a horse to be ridden to hounds has meant that a recognizable "type" has evolved.

Certainly, in terms of appearance, nowhere are the qualities of a hunter

better seen than in the show hunter classes held at major horse shows, though most of the horses in the show ring will never have seen a pack of hounds. The obvious reason for this is that the merest scratch or blemish, which is a natural hazard of riding fast across country, would spell the end of their showing career and dramatically reduce their value. Nevertheless, they must display those characteristics of conformation, movement, and, as far as can be judged in cold blood, temperament and good manners, that are most sought in a hunter. In some instances, they will have more "breeding" than is necessary for many hunters – show hunters are nearly always near-Thoroughbreds –

ABOVE: *Hunting is becoming more popular in the United States. This is the Westhills Hunt in California.*

but, particularly for people not wholly familiar with the look of a good hunter, it pays to spend some time visiting the major shows to assess the general qualities displayed by these horses.

Before you buy a hunter, you should consider the type of hunting you will be doing and the type of country you are going to be riding across. These factors will inevitably influence your decision as to what sort of horse to buy.

For drag-, bloodhound and stag-hunting, you need a horse with a considerable amount of blood

(Thoroughbred) in its breeding. All three sports are speedy and the latter two involve a considerable amount of jumping as well. This means that a horse of tremendous stamina is required. Open grassland or moor country calls for the same sort of horse, because again, the day's sport is likely to be fast. A lightweight, racing type is ideal.

In heavily enclosed, highly cultivated farming country, on the other hand, a shorter-striding, more strongly built horse, with a rather more placid temperament, would be infinitely more suitable. The finer legs of the Thoroughbred horse are not as suited to traveling across heavy plowed land, nor would its more temperamental nature be useful when it is asked to wait its turn at an enclosed, narrow jump, or while hounds are taking some time to come through a covert. Equally, speed is not such an important factor, for, in such country, hounds tend to run for shorter bursts at a time.

Horses required to follow a pack of harriers must also possess excellent speed, stamina, and good jumping ability. Although, like stag-hunting, it is conducted in moor country, a follower with harriers generally has to be able to jump to keep up with the pack. One mounted follower said it was not uncommon to jump more than 100 stone walls in a day with the pack he was following. You will therefore need a horse that is a good jumper and a good stayer.

SOME GOOD AND BAD CONFORMATION POINTS

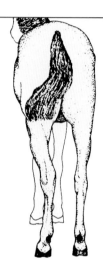

1 Good conformation. There is a straight line running from the point of the buttocks down through the hock to the hoof.

2 The hocks are turning in – a condition known as cow hocks. For extreme hard work, such a horse should be avoided.

3 The horse is bowlegged, that is, his hocks turn out and his toes turn in. This puts a great strain on the ligaments of the legs.

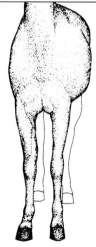

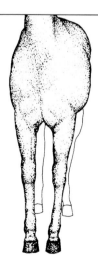

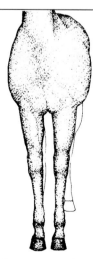

4 Again, good conformation. A straight line runs from the point of the shoulder, through the knee to the hoof.

5 The horse is pigeon-toed, which makes him prone to stumbling.

6 The front legs are set too close together, making less room for the heart and lungs. The horse is also likely to brush one leg against another, which soon causes injury.

HARNESS RACING

This has been a popular sport in the U.S. since 1790, at which time it was primarily an amateur affair. Since then it has become one of the most exhilirating sports ever, and popular all over Europe, Australia, South Africa and the United States. Selective-breeding in America has led to the rearing of the fastest harness racing breed in the world, the Standardbred, renowned not only for its speed but also for its stamina, keenness, and calm temperament.

Harness racing can involve either trotting or pacing. In both, the horse pulls a light two-wheeled vehicle, known as a sulky and is guided by a driver. The object is to race for one mile in the shortest possible time. In 1879, the Standard

was set in the American Trotting Register at two minutes and 30 seconds for trotters, and two minutes 25 seconds for pacers. Now speeds have reached as high as 32mph (51km/h), with record times of well under two minutes.

Trotters move diagonally, in a conventional trot. Pacers move laterally, that is, the legs on the same side move together. Pacing is a natural gait, and horses that are born with a talent for pacing are trained for it from an early age. Many pacers wear hobbles, which synchronize their strides and prevent them from breaking the trot.

Harness racing is a highly specialized sport and it involves quite specific equipment. A breast harness secures the horse to the

ABOVE: A pacer in the renowned Red Mile race. The movement is lateral, with the legs on the same side of the body moving together. They are generally slightly faster than trotters.

sulky. It consists of a breast collar, traces, saddle and girth, plus a bridle fitted with long driving reins. The collar fits round the horse's blind, named after the notable American trainer and driver, Thomas W Murphy. Basically it is a breast, while the traces run back from each side of the breast collar and are fastened to the sulky, which the animal draws, and in which the jockey sits.

Speed is the object of this sport and it is of paramount importance that a horse holds its head absolutely straight in front of him in order to

ABOVE: *The hobble is used to synchronize the strides of the pacer, and therefore to prevent the horse from breaking its trot.*

achieve his maximum. Not only that, but he will almost certainly strike into and injure himself when moving at a racing pace. There are various devices that have been designed to cope with the problem of the horse who persistently leans to one side or another. One of the best known methods is the Murphy blind, named after the notable american trainer and driver, Thomas W Murphy. Basically it is a piece of stiff leather that fits onto the cheekpiece of the bridle, and that is shaped so that it cups inwards slightly at the front of the eye. The principle of operation is simple; with a horse who turns his head to the left, the blind is fitted on the right, so that too acute a turn of the head will bring the blind in front of the right eye, thus obscuring his vision. A horse quickly appreciates that if he keeps his head straight he can see straight ahead perfectly well, but that if he turns it his vision will become obscured.

Other items in common use on harness horses, particularly pacers, are shadow rolls and hobbles. The

shadow roll is basically a sheepskin-covered noseband, fitted so that a horse can see straight ahead but cannot look down at the ground immediately in front of him. It is used on horses that tend to shy at shadows, marks on the tracks, bits of paper, and so on. Pacers are particularly prone to spookiness, something which it is believed, is the result of their wearing hobbles. Because they do not have free use of their legs they seem to be much more fearful of stepping in a hole or tripping up than other horses are. As a result, they are likely to shy violently at real or imagined objects in the ground, something which is extremely dangerous when racing alongside other sulkies.

Hobbles are adjustable straps fitted with padded loops at each end. One loop encircles the front leg and the other the hindleg on the same side of the animal. At the front and back of each leg loop is a vertical

strap which fits onto a set of four hobble hangers, which all fasten onto the back strap. The fitting of the hobble is extremely important; if the straps are too tight, they will prevent the horse extending properly and will tire the legs. If they are too loose a horse that is used to them may start to roll about in his gait as he seeks their support.

Harness racing is an incredibly fast sport, and can be extremely dangerous for both horse and driver. A horse can reach high speeds, and can injure itself in the process. Trotters often strike their elbows with their feet, and pacers the inside of their knees with the oppostie forefoot. To offer protection to the horse in these circumstances, there are various boots designed to offer protection to the shins, ankles, and coronets.

BELOW: *The horses manage to reach speeds of up to 30mph (50km/h), and there are times in a race when sulkies will be driven very close to each other. It is vital that the horse looks straight ahead at all times and is not distracted by anything, which could lead it to falter, or cause collision.*

MOUNTED GAMES

ABOVE: *Jousting competitions could be described as the medieval knight's answer to mounted games.*

More than with any other sport included in this book, a mention of mounted games conjures up widely differing views of equestrian activity to varying groups of people. To cavalry or mounted police, they are likely to mean skill-at-arms. This is a term that encompasses a number of "games," often ridden competitively, but principally designed to perfect the riders' use of weapons when mounted, at the same time as training their horses in obedience and combat tactics. The cowboys and girls who spend their days in the saddle, riding the range and working with cattle, will think of the competitive games that form part of the day's events at a rodeo. Members of riding clubs and similar organizations will associate the words with various noncompetitive, activity rides, such as "mock hunts" in which riders take the place of the fox and hounds, while others act the parts of the hunt staff and members of the field. The words could just as easily refer to the various informal permutations of the game of polo – such as cushion polo or paddock polo; to a type of mounted lacrosse known as *tshenkburti* in some

countries and polocrosse in others; to mounted paperchases and treasure hunts, or to any number of the hundreds of horseback games that have been devised throughout the centuries all over the world as tests of skill and obedience or as pleasurable relaxation.

International gymkhana

For many young riders, mounted games mean the competitive games that make up a major part of the schedule at gymkhanas or local horse shows. In Britain, the best-known version of these is the Pony Club Mounted Games Championship, known widely as the Prince Philip Cup Games. This is an eagerly contested team event that takes place throughout the United Kingdom between Pony Club branches during the Easter and summer holidays, culminating in the exciting finals that take place annually at the Horse of the Year Show, London. They were first staged there in 1957, when 45 branches of the Pony Club competed. From these beginnings, the idea spread to Europe. British Pony Club teams have demonstrated these mounted games in Holland, Germany, and France and youngsters from Belgium have begun to compete in mounted games internationally by sending a team to participate in an international event held at the Windsor Horse Show.

Pony Clubs in Canada and the USA also run a mounted games competition, known in Canada as the Prince Philip Cup Games, but in the USA as the Mounted Games of the United States. Similar to the British, who have zone and regional rounds to qualify for the national

finals at Wembley, teams in the USA and Canada compete in various qualifying rounds in the hopes of getting to their national finals. These tend to be a whole day affair and, though competition is fierce, the principal emphasis is on fun. At about the same time each year there is also an event known as an "international visit" in which teams from Canada, the USA, and Britain compete against one another. The venue is rotated among the three countries and the competition is open to children who are under 16 years old on May 1st of that year. (Competitors in the Prince Philip Cup Games competition in the UK must be under 15 years old on May 1st.) Australia and New Zealand, too, have their own versions.

Origins and value

Whatever the interpretation of the words – whether the game is a tent-pegging competition at the British Royal Tournament, a barrel race at a Texan rodeo, a bending race at a gymkhana, or a furious game of *buzkashi* played in the blazing deserts of Turkestan, the origins of all mounted games are the same. In their myriad forms of today, each one has evolved from the "games" and exercises practiced by mounted soldiers throughout the ages to prepare them for the skills and disciplines they would need to fight their enemy from horseback. The great Greek horsemaster, Xenophon, who lived in the 5th century BC, is known to have included such games as part of the training of his soldiers, and every cavalryman since has been similarly trained.

One of the great values of mounted games lies in the precision and accuracy they demand from both horse and rider. A horse must be instantly obedient – stopping, starting, and turning at a moment's notice. Because the rider is involved so often in some additional activity – such as spearing some object with a lance, putting a flag in a bucket, or passing a baton to a team member – the horse also must be both unperturbed by what his rider is doing and yet ready to respond to a command the instant it is given. For young riders, mounted games are particularly valuable, since they serve as an introduction to more serious, competitive equestrian riding, teaching competition tactics, behavior and etiquette.

Few mounted games, however, have the professionalism, and thus the seriousness, of competition associated with such sports as eventing or show jumping. This could be attributed to the fact that they have traditionally been played for relaxation and enjoyment – as lighthearted competition among friendly rivals and a way of promoting team spirit.

The mounted games horse

Because of the nature of mounted games – that is, their somewhat lighthearted, competitive element or their general use as part of more complex overall training – few riders seek to buy a horse or pony solely to take part in them. A cavalry horse, for example, would not be purchased just for its prowess in mock battles or a tent-pegging competition; nor would a cowboy buy a pony simply because it showed remarkable aptitude in the rodeo mounted games. In both cases the suitability of the animal for its daily work is more important. In addition, almost any horse or pony can be trained to participate in some form of mounted games, provided that its rider is sympathetic, rides well and knows what he or she is trying to achieve. The chairman of the organizing committee of the

British Prince Philip Mounted Games, Norman Patrick, said, for instance, that, even given the highly competitive nature of these games today, success is within range of any pony and child, *provided that they train hard enough*. Having said this, however, there are some valid points to look for if choosing a pony.

Selecting your mount

The size of horse depends on the rider. Never be tempted to buy a pony that is too small for you because you happen to know he is a good gymkhana pony. It is not fair on the pony and, in another year's time, the chances are you will be too big to ride him at all. There is also now a rule in the Prince Philip Mounted Games which states that no child who weighs more than 8½ stone (54kg) when dressed in riding kit may ride a pony of 12.2hh or under.

Points to bear in mind about size are that smaller ponies are useful for games that involve, for example, bending down and dropping something into a bucket – it is not so far to reach, so the margin for error is less – or when vaulting on and off your mount quickly is involved. On the other hand, such ponies will often have a short stride, which means that they cannot move as fast as their longer-striding, taller rivals, and speed is undoubtedly an important factor in some games. It is sensible, therefore, simply to look for a pony that is the right size for you to ride on an everyday basis.

The age of the pony is another factor. No pony under four years old is allowed to compete in the Prince Philip Games; indeed, if you buy a pony of this age, you will need to be a good enough rider to train him

LEFT: *Barrel-racing is the rodeo equivalent of a gymkhana bending race. Note the angle of the pony's legs to the ground; he is well-prepared to make a very tight turn around the barrel.*

FITNESS

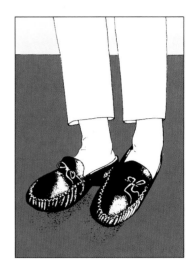

ABOVE: *It is essential that, as a rider, you are as fit as your mount. You need to exercise regularly to maintain a high level of suppleness and agility.*

completely in the art of mounted games. Undue excitement – which is certainly part of any competition, can easily cause him to "blow up" – that is, become excitable and uncontrollable – whereas an older pony of six years or more is likely to be more settled and to act more maturely and sensibly. Many Pony Club mounted games trainers consider the ideal age for a Prince Philip pony is 10 to 12, by which time such animals have considerable experience, are thoroughly used to life, and less likely to be upset by the noises and distractions of a mounted games event.

Conformation is important. A horse or pony that is "well put together" will always perform better than one with poor conformation. The ideal animal will be short, rather than long, in the back, as this makes for greater agility of movement. The neck should be slender rather than thick, and the head small but well-proportioned to the rest of the body. A horse uses his head and neck to balance himself, particularly when moving at speed, and a large head and thick neck tends to lead to clumsier movement. The legs are important, as they have to take considerable strain when, for example, a horse is asked to turn and

stop at a moment's notice. They should be clean, strong, and unblemished, but not excessively fine or slender. Make sure the pastern slopes forward correctly; an upright pastern has a less efficient shock-absorbing action.

Smoothness and evenness of paces are things to look for and you should check these by watching the horse or pony moving toward you, past you, and away from you, as well as by the feel of the movement when you are on his back. Check the three major paces – walk, trot, and canter. They should look and feel evenly balanced and rhythmic. Similarly, the animal should feel well-balanced as you ride him around corners and in a circle. It is extremely important for a mounted games pony to answer his rider's leg commands instantly. If you feel he does not do so as you are trying him, consider whether he is sufficiently young or amenable to respond to training.

The temperament is of paramount importance. A headstrong or bad-tempered animal – one that lays his ears back as you approach him or looks generally bad-tempered – is to be avoided. He will be a menace to you and others. The ideal temperament is one that lies midway between being very high-

spirited and hot-blooded, and very docile and too easygoing. An excessively high-spirited horse or pony could prove too excitable for competition, but a very docile one might be too lazy when it matters. Look for an animal, therefore, with a certain amount of spirit and certainly one which takes an interest in what he is doing and what is going on around him.

ABOVE: *The sack race is a perennial favorite, nearly always present on mounted games schedules at gymkhanas.*

Specialist training

Training a pony to compete in gymkhana-type mounted games will be both more fun and more effective if you can collaborate with some friends. This not only introduces the important element of competition, but also means you can help and correct each other as you progress. If a riding club runs selection sessions, go along to these. Even if you do not get picked for the team, participation in the sessions will give you a basic idea of how to train yourself and your pony.

There are two important rules connected with the training of a pony for mounted games. First, a rider does not need a whip or stick. A pony should never be punished by hitting it, nor should a stick be used as a means to encourage him to go faster. Most competitive events ban the use of whips and spurs. Second, never jerk your pony in the mouth. In the excitement of the moment, when trying to turn sharply or stop quickly, an enthusiastic rider is quite capable of forgetting the proper use of the hands, so remember this warning and make sure this never happens.

Among the most important things for a gymkhana pony to be taught are to lead readily, either alongside of another pony and rider or while you run beside him, and to stand still while you mount or when another pony is coming full tilt toward you. This will be necessary in team relay races or some pairs competitions, in which you must receive a baton from a fellow rider before moving off.

Most ponies will lead readily in hand, but you should practice this at a walk and a trot while you walk or run beside the pony. Put him in a halter the first few times; if you haul on the reins in an attempt to lead him, he will naturally shy backward to try to get away from the bit banging on his teeth. If the pony does hang back, get a friend to stand behind him and urge him

forward. The persuasion should be calm and gentle; a pat on the hindquarters and an encouraging click with the tongue – not a heavy slap, waving of arms, and shouting.

It is basic good manners that a pony should stand still while you mount; and even top competition ponies, who know that they have to move off at a gallop a moment later, should stand still until their riders are in the saddle and give the signal to move. If your pony develops the habit of moving as you mount, get off and stand by his side, controlling him with the reins. Make him stand still until you are in the saddle, and then keep him still for a minute or two. Make sure that the aids you give when you ask him to move off are clear and definite. He must learn that sometimes he has to stand still for the start of the race and at other times you will want him to move off immediately.

Keeping a pony still while another is galloping flat out toward him may take a little patience on your part. The first few times they encounter this, most ponies will jump to one side or whip around.

ABOVE: *Racing close together, side by side, is a feature of some mounted games. Practice doing this with a friend; it can be a little unnerving for a pony until he gets used to it.*

Initially, therefore, ask someone to hold your pony's head in order to keep him still, while you pat his neck and talk to him. He will soon learn to stand still. This is important in a relay race, for instance; you will be concentrating on receiving the baton and want the pony to stand still without having to think about it.

It is also a good idea to try to train your pony to understand and respond to your voice. If you can calm him at the start of a race by talking to him quietly or slow him down when you want to with the command "steady" or "whoa," you will find this an immense help. Get into the habit, therefore, of using such commands each time you slow down or come to a halt – making sure your voice is always slow, even, and gentle. It is the tone and pitch of your voice he will come to recognize – "whoa" screamed harshly will have the opposite effect to the one intended.

Turning and halting

Good gymkhana ponies must be able to turn sharply and to stop instantly. The ability of a pony to turn sharply – around a post, for example – depends not only on his flexibility and agility, but also on the rider's ability to use his legs correctly. Practice doing tight turns around a post at a walk and then a trot; the aids are the same for any turn – slight pressure on the rein in the direction you want the pony to turn, pressure with the outside leg – the leg on the opposite side to the turn – behind the girth and pressure with the inside leg slightly further forward. Make sure your hand does not move; it is a common sight to see a rider, when asking for a tight turn, with his hand pulled back into his stomach. Turning sharply back on your own tracks differs only from turning a 90-degree bend in that you keep on using your legs and keep on exerting slight pressure on the rein through almost 180 degrees. When you are making smooth, sharp turns at a walk and trot, you can begin practicing a canter.

Stopping quickly should be practiced carefully, for it is very important to stop correctly as well as quickly – that is, with the pony's legs well beneath him and his head properly positioned. Another common sight is a rider hauling sharply on the reins in such a way that the poor pony does indeed stop quickly, but with his legs askew and his head flung into the air in an attempt to avoid the pressure on his mouth. Ask your pony to stop from a trot by closing your legs against his sides – this will bring his hind legs under him – and resisting with your hands as he moves forward. When he is stopping smoothly, ask for the same thing from a canter. If anything, this will be easier; you are likely to be leaning forward slightly in the saddle, as you have been urging him forward. So, when you want to stop, sit back in the saddle, apply the same closing action with your legs, and once more resist with your hands. If you have been riding correctly on a loose rein, the shift of your weight in the saddle will already indicate that you want to slow down.

BELOW: *It is never too early to start! Even if you are not yet ready to compete in many mounted games, it is a good idea to go to some gymkhanas so that both you and your pony get used to the procedure.*

Rehearsing the game

Undoubtedly the best way to train for individual games is to ride through the actual procedure for them. Initially, it is best to do this quietly on your own at a walk, progressing slowly to a trot, and finally to a canter. Do not start even friendly competition with others until both you and your pony are proficient.

Always work through the movements of a game from start to finish, and then go through them again until you get them right. You must make sure that your pony is performing correctly; that he is bending his body properly (if this is required); that his head is properly positioned; that he is moving forward in a well-balanced fashion, and that he is listening to you all the time. Be patient, since no horse or pony can be expected to know what he must do in a game until he has been taught, and it may take him a little while to understand what exactly is required, particularly if you are asking him to learn all sorts of different games. If you are experiencing difficulty in some particular aspect of a game, do not go over and over it until both you and your pony are thoroughly frustrated. Leave it and come back to it, preferably after you have asked advice.

You cannot expect your pony to be brilliant at all gymkhana games. In some, the emphasis is on speed; in others, it is very much on the "game" element. Few ponies are good at both, and also there are some games that individual animals simply do not like. If, however, your pony appears not to perform well in a particular race for no apparent reason, look at what you are doing or not doing, rather than instantly blaming him. Are you riding correctly, giving clear leg aids, and using your hands sympathetically? Particularly if you

were racing against others, think how you used your voice. The excitement of the race may have made you alter the tone or pitch, and ponies are sensitive to this.

Intersperse gymkhana training sessions with other types of riding, such as hacking or jumping. Ponies well-versed in games get to know them so well that practicing the same games over and over again will make them bored, stale, and ultimately uninterested. Once you have mastered the technique of a particular game, there really is not much point in running through it endlessly; turn to some other game instead.

Bear in mind, too, that it is a good idea to limber up a pony gently before embarking on a training session. Give him a quarter of an hour going through a few gentle school paces, or try some light interval training. In this, you walk your pony for three minutes, rest him for three minutes, trot for three minutes, rest for three minutes, trot again, rest again, and then canter for three minutes. Do this on either rein and concentrate on what you are doing. As well as making him obedient it is a great help in getting a horse or pony fit. It will also settle him a little before practicing the games and make him cooperative from the start.

Preparing for competition

Practicing mounted games in a paddock with a few friends and competing in a gymkhana against other keen contestants can be two very different things, unless you are well-prepared.

If your pony is not already a seasoned competitor, you should consider the effect that going to even a small competition could have on him. He is going to be taken to a strange place (probably in a box or trailer) alive with noise and bustle. There will be numbers of cars and horse boxes, and endless horses,

ponies, and people milling around. There will be tents and marquees, roped-off competition rings and, highly likely, a voice booming over a public address system. This will naturally be quite a shock to a pony or horse who has not experienced it before. In addition, you are going to ask him to remain calm for part of the time and race at top speed at other times.

Do all that you can to acquaint a pony with similar conditions before expecting too much of him. It would be a help to take him to a small show where you are not competing, just to get him used to the many distractions.

Gymkhana games take place within an arena or marked-off area of some kind. If your pony is not used to entering an arena, he may find this an unnerving experience, and if he fails to go in quickly, you may be eliminated from an event. Try, therefore, to get him used to riding in and out of a roped-off area. Walk him in and out of it several times without running any races, so that he does not associate the arena with racing. If this happens, he will start to get excited the minute you enter.

RODEO

odeo (the Spanish word for a cattle ring) is a survival from frontier days in the USA. Then, the trail gangs held informal competitions after a cattle drive to demonstrate their skills in the arts of roping, bareback bronco riding, steer wrestling, steer and calf roping, bull riding, and so on. From these beginnings, the spectacle of rodeo was born.

As far as is known, the first public rodeo was held on 4 July, Independence Day, 1866 in Arizona. By the turn of the century, the Wild West show had come into being, adding an element of circus and carnival to the exhibition, with demonstrations of fancy shooting, riding, and chuck wagon racing. Public interest is now on an international scale. In the USA today, according to the Rodeo Cowboys' Association, more than 500 professional rodeos are held annually, as well as the hundreds of amateur contests. The total audience runs into the millions, while prize money, too, can be considerable. For its part, the Australian Rough Riders Association has a membership of over 12,000 and organizes rodeos in every state. The National Finals Rodeo is the highlight of the Australian rodeo season.

The events

There are five traditional rodeo events; calf roping; steer wrestling; and bronco, bull, and saddle bronco riding. All of them require considerable skill and nerve from both horse and rider. In calf roping, for instance, a young calf is released into the arena from a chute; the cowboy's task is to gallop after it and lasso the animal. The horse then comes to a sliding halt and backs

BELOW: *Calf roping is one of the rodeo events that is closest to genuine cowboy life and is an integral part of a western horse's basic training.*

away to maintain the tension on the lariat, which is secured to the saddle horn. This is all part of a good western horse's basic training. The cowboy dismounts, runs over to the calf, turns it on its side, and quickly ties three of its legs together. The contestant with the fastest time wins.

Another test of skill, in which the risk of injury can be considerable, is bronco riding. In saddle bronco riding, the bronco wears a saddle and the rider tries to stay seated in it for a minimum of ten seconds. His only security is a halter rope, which he holds with one hand. In bareback bronco riding, the saddle is replaced by a surcingle strap with a leather handhold. The stipulated minimum time is eight seconds.

In both events, the contestants score points up to a maximum of twenty-five, awarded by two judges on the basis of the riders' skill and the horse's wildness. To increase this, the rules stipulate that the animal must be spurred, on its shoulders, as it leaps from the chute, while hard spurring during the ride wins bonus points.

Women riders, too, feature on the rodeo circuits; the Girls' Rodeo Association in the USA was formed as early as 1948. Rewards for successful riders are considerable, but expenses are also high. Competitors have to pay their own entry fees, and there is also the cost of traveling the thousands of miles involved to be taken into consideration. There is also the

ABOVE: *Ten seconds is a very long time to be sitting on a horse when it is bucking as much as this and you have no saddle! Many contestants find it very difficult to stay on.*

constant risk of injury – or even death.

This is one of the reasons for the employment of clowns and pick-up riders at rodeos. As well as entertaining the crowds, part of the clowns' task is to attract the attention of a bull, or bronco so that a fallen rider can be brought to safety. Pick-up riders, as their name implies, have the job of helping successful competitors off their mounts by riding alongside and lifting them to safety.

Hazers, on the other hand, have a less humanitarian role. Their task is to gallop alongside the bull in bull wrestling events to make sure that it keeps to a reasonably straight line. In some ways, their role is not unlike that of the picadors' in the bull ring, just as rodeo itself, particularly in its display and spectacle, has some links with bull fighting. The picadors' task is to bait the bull in preparation for the toreador and to help ensure his safety. Traditionally, toreadors fought on foot, the picadors being mounted. Recently, however, mounted bullfighting has gained considerably in popularity.

Cutting horses also compete in rodeos, though they have their own competitions and events. These are extremely popular in the USA and Australia; in the latter country, such was public demand that the National Cutting Horse Association was formed in 1972 to organize the sport. Watching the horse at work is almost like watching a sheepdog, as it cuts out a steer apparently without assistance from its rider.

ABOVE: *The prospect of death is not inevitable for the hazers, who can sometimes get dangerously close to the bull.*

BELOW: *Cutting is a large part of farm and ranch life. In America the cowboys use cutting horses to herd their cattle together.*

POLO

ABOVE: *A 17th-century version of polo played in provincial Mughal. It was not until the 19th century that the game was discovered by the western world.*

Polo is a tough, high-spirited, highly skilled, and extremely fast sport – in fact, it is the fastest team game in the world. Just when and where it was first played is not entirely certain; we know that a version of the game was being played in ancient Persia about 2,500 years ago, but how closely this resembled the modern game is open to conjecture.

Polo as it is known today was first introduced to the western world from India by cavalry officers in the middle of the 19th century. These officers had seen versions of the game played on tough little native Indian ponies, with the local village streets acting as the playing grounds. The rules varied from place to place, as did the number of players – this, it seems, largely depended on how many people wanted to join in. Nevertheless, the officers realized the potential value of the game, both as a relaxation

from their duties and as a valuable means of training horses and young officers. Hence, they began to play it among themselves, bringing it back to Britain around the end of the 1860s.

From there, the game quickly spread to the Continent and across the Atlantic to the USA, where today it has an impressive following. Gradually the rules became ordered and standardized internationally. In the first matches, the rules stated that the game should be played on ponies standing no higher than 13.3hh (although those used were often considerably smaller). Soon a new height limit of 14.2hh was imposed, until it was realized that even this was impractical and the restriction was abolished altogether. Nowadays most people ride "ponies" of about 15hh–15.2hh and this is generally thought to be the ideal height.

Polo is now played worldwide from Australia, New Zealand, Hong Kong, Singapore, and Malaysia to India, Pakistan, South Africa, Nigeria, Malta and Cyprus, the USA, Argentina, Jamaica, and Barbados. It has a following, too, in France and Germany, as well as in Britain. The game most common to all these countries is regulation or standard polo, but there are other, recognized, modifications of the game, of which paddock and arena polo are probably among the most widely played. These versions have only three players per team (as opposed to four in regulation polo) and are played in a considerably smaller area. Arena polo is often played in an indoor stadium or school and is popular in the USA.

Such modifications of polo are generally ridden at a rather less furious pace and often serve as an introduction to regulation polo for many riders. Another way in which young riders are finding their way into the sport is through the Pony

Club. Many branches intersperse games of paddock polo with rallies and other events, and there is also a recognized game of Pony Club polo. The rules for paddock and Pony Club polo are based on the official rules.

The game

The game of polo varies very slightly depending on whether you are concerned with paddock or arena polo (which may be played indoors or out), Pony Club polo, or regulation outdoor (standard) polo. Furthermore, different countries have their own regional modifications in rules and play. However, all the variations are very slight and generally speaking the principles are similar whatever game you are playing.

In paddock or arena polo, as mentioned, there are only three players a side; in other polo games there are four. Numbers one and two are the attacking players or the forwards; number three – the half-back – is described as the pivot of the team in that he is both attacking and defensive, while number four, known generally as the back, is the principal defensive player. (When there are only three players, numbers one and two are combined.) Each player holds a specially designed mallet or stick, with which he hits the ball. It has a long, cane shaft and a cigar-shaped head. The ball in standard polo is made of willow or bamboo root and is about 3¼in (8cm) in diameter. In paddock and arena polo, the ball is a little larger and softer, made either of leather (inflated like a football) or plastic. The size of the playing area varies from 300yds (279m) long by 160yds (146m) for standard polo, to an area about a third of this size for arena polo. For standard polo, the width is increased to 200yds (183m) if there are no marking boards bounding the area.

In all kinds of polo, the object of the game is to score goals, and the winning team will be the one with the greatest number of goals to their credit at the end of the match. Goals are scored by players hitting the ball through the goalposts, which are situated at either end of the playing area in the center of the short sides of the field. A match is divided into periods of play known as chukkers. The number and length of chukkers depends on the type of polo and the importance of the match. As examples, tournaments of top-class standard polo are usually divided into six chukkers of 10 or 7½

minutes each, while paddock or Pony Club polo may comprise only two chukkers of six minutes each. Chukkers are separated by intervals of three to five minutes. (The actual time of a chukker may be longer than that specified for the match, as the clock is stopped each time a foul is committed until play resumes.) Play is divided into these comparatively short spells because of the extremely fast pace at which polo is conducted. It would be inhuman to ask a horse to continue for longer without a break – in big matches a player will ride two or three ponies.

Players line up in formation at the start of a game and the referee or

umpire throws the ball between the lines. The play takes place mainly up and down the field and the players mark each other while always attempting to move the ball in the direction of their scoring goal. Player number one will generally aim to be nearest to the goalposts through which his team will score, with the other players spread down the field so that number four stays closest to the end he is defending. Number one therefore plays opposite the other team's number four and so on.

The ball is passed between players towards the opposing goal, while the other team tries constantly to intercept it. When a goal is scored,

THE OFF-SIDE BACKHANDER

1 Shift the grip to reverse.

2 Move your right arm forward so the stick lies almost parallel with the pony's back over your left shoulder.

3 Swing the stick in a backward arc, rotating your shoulders from front to back. Continue to watch the ball.

4 Continue to rotate your shoulders after hitting the ball, until they are almost in line with the pony's spine.

THE OFF-SIDE FOREHANDER

1 Tighten your grip on the stick and raise your hand to the right.

2 Begin to turn your body to the left, bringing the stick forward.

3 Swing the stick in a backward arc by rotating your shoulders. Continue to look at the ball and lean over to the left.

4 Follow through by continuing to rotate your shoulders and bringing your hand up level with your shoulder.

the teams change ends and play begins again. If a ball goes over the sidelines, the teams line up as for the opening of play 5yds (4.5m) from the side and the umpire throws it between them again. If the attacking side hit the ball across the back line without scoring a goal, the defending side hit it back into play from that position. If the defending side hit the ball across the back line, the attacking side are given a free hit from a prescribed distance away from their goal. If the ball is still in play at the end of a chukker, it is thrown into the teams, who still line up in the usual way, but the ball is thrown toward the side where it had been when play stopped.

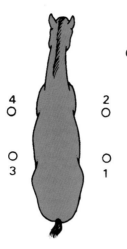

4 ○	2 ○
3 ○	1 ○

1 Off-side backhander: the ball is positioned just behind the girth on the off-side.

2 Off-side forehander: the ball is positioned opposite the point of the off-side shoulder.

3 Near-side backhander: the ball is positioned just behind the girth on the nearhand side.

4 Near-side forehander: the ball is positioned opposite the point of the nearside shoulder.

LEFT: *Diagram showing the position of the ball when making each of the strokes outlined below.*

BELOW: *The ball is struck using the long side of the stick head.*

THE NEAR-SIDE BACKHANDER

1 Swing the stick backward, by turning your right shoulder back so that it points towards the pony's tail.

2 Turn your wrist outward and pause for a moment at the top of the swing to help you time the stroke correctly. Your left shoulder should be pointing forward and you should be leaning to the right.

3 Keep your head still, looking down at the ball, and bring your right shoulder forward, keeping your arm straight as you hit the ball.

4 Let the stick swing forward to follow the stroke through.

THE NEAR-SIDE FOREHANDER

1 Shift your grip to reverse and begin to take your stick to the left side.

2 Take your arm right back, turning your body from the hips.

3 Swing the stick in a forward arc, keeping your hips still.

4 Allow the stick to follow through, so that you end up leaning well forward in the saddle.

Riding-off and right of way

Players attempt to intercept the line of the ball or stop a member of the opposition getting it by riding-off their opponents, leaning against them, and by knocking or hooking sticks. Riding-off means riding alongside of an opponent, using your pony to push his off the line of the ball, while hooking an opponent's stick is done by blocking the line of his swing as he nears the lowest point, thus catching his stick before it reaches the ball. A string of regulations surround these and other maneuvers, all designed to safeguard the players and their mounts. Any infringement will lead to a foul being declared, and a penalty awarded to the opposition. Penalties generally take the form of a free hit at the goal taken from a designated place down the field, the distance depending on the severity of the foul.

The rules surrounding right of way are the most imporant to understand fully, for infringement of these generally accounts for the greatest number of fouls in a match. The official rules go into the various possibilities in some detail, but, in essence, the player who is following the exact line of the ball – that is, the direction in which it is traveling, so it will be on his off-side, or is at the smallest angle to that line – has the right of way. No other players may cross this line if, by so doing, there would be a risk of collision, unless the player with the right of way were to check his speed, although they can attempt to ride off that player. There are exceptions – for instance, when two players coming from directly opposite directions will have equal right of way – so players must go deeper into this aspect before entering the game.

Probably the most important overall point to remember about playing polo – whatever the type – is that it is a team game, in which players must continually support one another and play in coordination. Experienced players will have team tactics and strategies calculated; less experienced players do better to adopt a policy of always backing up other team members, so that if someone misses the ball, there is a member of his or her team not far behind to take over play.

The polo pony

The polo mount is always known as a pony, even though in many instances he will stand higher than 14.2hh. The term "pony" stems from the time when the official rules imposed a height restriction of no more than 14.2hh for all horses to be ridden in polo matches.

The type and size of mount you will be looking for will depend mainly on the type of class of polo you intend to follow and also, up to a point, in which country you are going to play. Most polo-playing countries produce a "polo pony" of some sort and, by and large, for an average-standard game (i.e. not high-level tournament play) this is usually the best to buy, as it is likely to be best suited to the prevailing conditions.

If your involvement with polo is to be playing Pony Club polo with other Pony Club members, an ordinary pony – preferably one that excels at mounted games – is ideal. His participation in the latter will have developed his agility and his ability to stop and start quickly and turn tightly. In addition, he is attentive to you so that he responds quickly to your commands. All he will need, therefore, is a little additional training to get him used to the stick swinging by his side and to make him familiar with the most common techniques.

It may be, however, that you want to look for a pony just for polo. Although there are certain guidelines you can follow, your choice will again largely be governed by the type of polo and your experience and standard as a player, as well as by how much you can afford. If you are a complete beginner, it is best to start playing on an experienced polo pony – one who can teach you a thing or two about the game. Ask the advice of an experienced polo player and a good judge of horseflesh before making your choice. A trained polo pony will be at least six years old (his training could not have been completed before this time), ideally between 15hh and 15.2hh. He should be fast, with a smooth, low-galloping action, able to stop instantly from a gallop on command, willing to move off into a fast canter from a halt and with the ability to turn very tightly while moving fast. In addition he should be familiar with the "riding-off" technique and leaning against other horses while moving at a gallop, as well as being able to hold a straight line to the ball at whatever pace he is moving. A polo pony will be at his prime between eight and 10 years old and a good one of this age will be extremely expensive. One that is a little older, perhaps 12, would still have a good few years' playing in him (particularly in slower matches), would be more reasonably priced, and could still teach you a great deal about the game.

If you have some experience of the game, you may prefer to buy a pony that appears to have the potential to become a good polo mount and train it yourself. For top-class polo, it is generally considered almost essential to ride a Thoroughbred, as this is the breed that possesses the necessary turn of speed. Many people look for ex-racehorses – those that have been consistently beaten on the racetrack by faster horses, not those that have been rejected from racing because of persistent leg trouble. Providing a racehorse is sound (have him very thoroughly checked by a veterinarian), not too big, and has not been ruined during his racing career by having too much asked of him, he could be made into a superb polo pony by someone who

knew how to do so. And, as a reject from one sport, he may be comparatively inexpensive.

As mentioned above, the ideal height for a polo pony is about 15hh to 15.2hh. A taller pony than this will be less agile and will not maintain the desirable pony characteristics, such as a smooth, but not too long, galloping stride. The higher he is, too, the harder you will find it to hit the ball, as you will have further to lean down. If you are planning to train your own pony, you could look for a horse of about five years old; below this age he will not yet be physically mature enough to be subjected to the rigors of training for polo.

The conformation you would look for in a good, or potentially promising, polo pony is similar to that of an eventer – that is, an athletic appearance; not too long in the back; rounded, muscular hindquarters; well-rounded ribs; good depth of girth; sloping shoulders; shortish, well-proportioned neck; pleasing head with a kind eye; clean legs with a long, upright humerus (top part of the foreleg); a low-down stifle; well-formed hocks; short cannon bones, and sloping pasterns. The feet must be well and evenly shaped – the right size for the horse. He should look able to carry your weight easily, but not too clumsy or cumbersome.

Soundness is of paramount importance but, if you are buying an experienced polo pony, you will be very lucky to find one with perfectly clean legs. Take professional advice to ensure that any blemishes you see or feel on the legs will not affect the pony's action or performance.

Temperamentally, you want an animal that is intelligent, eager, and bold, but calm in character and always willing: but most important, he must be obedient.

Argentinian ponies are almost universally considered to be the best polo ponies in the world. Polo was first played in Argentina as a means of relaxation and friendly rivalry among the cowboys and ranch hands. The handy little cow ponies proved to be so well-suited to the game, that astute players began to breed the ponies more selectively, by crossing them with imported Thoroughbreds. The offspring were then trained in the techniques of polo-playing, and they are now widely exported to the polo-playing countries of the world.

GLOSSARY

"Against the Clock" A term used in show jumping competitions in which the final round is timed. The winner is the competitor who has the least number of penalty points combined with the fastest time over the course.

Aids Recognized signals used by a rider to pass instructions to his mount. *Artificial aids* include whips, spurs and ancillary items of tack used by a rider to assist him in giving aids. *Natural aids* are the rider's hands, legs, body or voice. *Diagonal aids* are aids in which opposite hands and legs are used simultaneously, ie the right rein is used with the left leg. *Lateral aids* are hand and leg aids given together on the same side.

Airs Above the Ground High school movements (qv) in which, at some stage, all four legs of the horse are off the ground.

Apron Strong, leather or hide apron worn by a farrier (qv) for protection when shoeing a horse. *Apron (side-skirted)* Name given to the "skirt" of the sidesaddle habit.

At Grass A horse that has been turned out in a paddock or field.

Balance Strap Leather strap attached to the off-side back of a sidesaddle which passes under the horse's belly and buckles to a strap on the front of the saddle. It is designed to prevent the saddle from slipping.

Bars (of mouth) Fleshy area between the front and back teeth on either side of a horse's mouth.

Bay A deep, rich, reddish-brown colored horse, with black mane, tail, and lower legs.

Bit Mouthpiece often made of metal, rubber, or vulcanite placed in the horse's mouth and kept in position by the bridle to aid the rider's control. *Curb bits* include any one of a number of bits, the mouthpieces of which vary in design but which include hooks on either side to which a curb chain or strap is attached. This lies in the horse's chin groove and gives the bit its characteristic leverage action. A *gag bit* is a particularly severe form of bit. It may be raised to a greater or lesser degree, thus affecting the severity of the bit. A *snaffle bit* is any one of a number of designs of bit that act on the corners or bars of the mouth. The bit takes only one pair of reins.

Bitless Bridle Bridle without bit. Control is achieved by concentrating pressure on the nose and chin groove. A *bosal* is a very simple bitless bridle, the term actually referring to the rawhide noseband which is its chief component. A *hackamore* is the most widely known type of bitless bridle.

Box, to To lead a horse into a horse box or trailer.

Break In, to Training the young horse to accept and respond to a rider on his back.

Broken Wind Permanent disability to a horse's respiratory system manifesting in a chronic, persistent, and rasping cough.

Brushing Striking of the inside hind or foreleg with its opposite. May lead to injury and lameness.

Cantle Extreme back ridge of a saddle.

Cavalletti Adjustable low, wooden jump used in the schooling of horse or rider in jumping.

Cavesson Either a simple noseband fitted to a bridle, or a more sophisticated piece of equipment worn by a horse when he is to be longed (qv). In the latter, it is sometimes referred to as a *breaking cavesson.*

Chaff Fine-cut hay mixed with a corn feed to provide bulk and prevent the horse from bolting the feed.

Chestnut An overall yellowish-brown coat, with the mane and tail possibly the same color.

Cob A type of horse characterized by its smallness and strong, thickset build.

Collected A horse that, while moving forward, indicates it is ready to respond to its rider and so is "collected" together: neck arched, hocks tucked well beneath it, and gait lively.

Colt A male, ungelded horse up to four years old.

Concussion Jarring of a horse's legs, usually caused by fast trotting on the road, or considerable hard work on hard ground. May result in swelling and lameness.

Counter Canter School movement in which the horse canters in a circle with the outside leg leading instead of the inside leg as usual.

Curb Chain Single or double link chain attached to the hooks of a curb bit and lying flat in a horse's chin groove.

Diagonals (left, right) A rider rides on the left or right diagonal at the trot depending on whether he rises as the horse's left or right foreleg moves forward. On a circle, the rider should always rise as the outside foreleg moves forward.

Disunited (canter) Canter in which the horse's legs are out of sequence.

Dorsal Stripe Darkened line (usually black) running along the horse's dorsal ridge.

Double Bridle Traditional bridle with two bits (snaffle and curb), giving the rider greater control than a bridle with one bit.

Draw Rein Severe form of control comprising of a rein attached at one end to the girth, which passes through the bit rings and back to the rider s hands.

Dressage The art of training a horse so that he is totally obedient and responsible to his rider, as well as agile and fluent in his performance.

Drop Noseband Noseband which buckles beneath the bit to prevent the horse from opening its mouth to "catch hold" of the bit, making it easier to ignore the rider's commands.

Dun Generally refers to a "yellow" coat with black mane, tail, legs, and dorsal stripe.

Emergency Grip Position used by sidesaddle riders when there is danger of being unhorsed.

Equitation The art of horseback riding and horsemanship.

Eventing Riding in a one- or three-day event, which combines dressage, cross-country, and show jumping.

Extension The lengthening of a horse's stride at any pace. It does not necessarily mean an increase in speed.

Farrier A skilled craftsman who shoes horses.

Fence A *combination fence* is a series of fences (usually three) in a show jumping course, placed to allow only one or two strides between each jump. A *double fence* is two fences in a show jumping course with the same requirement as the combination. A *drop fence* is an obstacle in which the landing side is considerably lower than the take-off side. A *spread fence* is one in which the main feature is the width rather than height, whereas an *upright* is designed to test a horse's ability to jump heights.

Fetlock (Joint) The lowest joint on a horse's legs.

Filly A female horse up to the age of four years old.

Foal A horse of either sex up to the age of one year old. Male foals are usually referred to as colt foals, females as filly foals.

Fodder Any type of foodstuff fed to horses.

Forehand Front part of the horse including the head, neck, shoulders, and forelegs.

Frog V-shaped leathery part found on the soles of a horse's feet which act as a shock absorber and as an antislip device.

Gait The paces at which a horse moves. Usually, a walk, trot, canter or gallop.

Galls Sores caused by ill-fitting saddlery.

Gamgee Gauze-covered cotton batting used beneath stable or exercise leg bandages for extra warmth or protection.

Gelding A castrated male horse.

Grackle Noseband Thin-strapped noseband with double straps buckling above and below the bit.

Grass Tips Half-moon-shaped shoes which cover only the toe area of the hoof. Used for horses turned out to grass to prevent the hoof from growing too quickly (which creates a falling hazard).

Gray Refers to any color horse from pure white to dark gray. Further described by such terms as "dapple-gray" (small iron-gray circles on a lighter background), "flea-bitten" (specks of gray on a white background), etc.

Groom Person who looks after the daily welfare of a horse.

Grooming Kit The various brushes and other tools used in cleaning a horse's coat.

Ground Line Pole or similar placed in front of a fence to help horse and rider judge the takeoff point.

Habit Traditional riding kit worn by sidesaddle riders.

Hack A type of horse characterized by its pleasing appearance, fine bone structure, good manners, and complete obedience to its rider's commands. Also a term used to describe going for a ride.

Half Pass Dressage movement performed on two tracks (qv) in which the horse moves forward and sideways simultaneously.

Half Volte A school movement in which a horse is asked to leave the track and perform a half-circle of a given diameter after which he rejoins the track to continue in the opposite direction.

Hand The recognized measurement used for determining the height of a horse or pony. A hand equals 4in (10cm).

Haynet Large net or bag made of rope designed to hold a horse's hay.

Head Lad Used in racing stables to describe the head groom – the one who has overall responsibility for the welfare and general condition of the horse.

High School The classical art of riding, in which the traditional advanced school or dressage figures are practiced.

Hock The joint in the center part of a horse's hind legs. Responsible for most of the horse's forward force.

Hoof Pick A small, metal implement with a pointed hook on one end, used to remove dirt, stones, etc. from a horse's hooves.

Horn The hard, insensitive, outer covering of the hoof.

Horse Box Self-propelled vehicle used for the transportation of horses.

Horsemanship The art of equitation or horseback riding.

Horsemastership The art of caring for and attending to all aspects of a horse's welfare, under all possible circumstances.

Hunter Any type of horse considered suitable to be ridden to the hounds.

Hunting Head The top of the two pommels found on a sidesaddle.

The hunting head is in a fixed position and supports the rider's right leg.

Impulsion Strong but controlled forward movement in a horse.

Indirect Rein The opposite rein to the direction in which a horse is turning. When giving an indirect rein aid, the instruction to turn comes by pressing the opposite rein against the horse's neck.

Inside Leg The leg or legs of rider or horse on the inside of any circle or track being described.

Irons Stirrup irons are metal items of tack attached to the saddle by the stirrup leathers to hold the rider's feet.

Jog Western-style riding term for trot. Also used in European style riding to describe a slow, somewhat shortened pace of trot.

Jumping Lane A narrow track, usually fenced on either side, in which a series of jumps are placed.

Keeper Small, leather loops found on the straps of a bridle, designed to contain the end of the strap after it has been buckled, giving a neat appearance.

Leading Leg The front leg at a canter or gallop that appears to be leading the leg sequence.

Leading Rein Long rein attached to the bit by which the horse may be led. Usually used in the early stages of being taught to ride.

Leaping Head The lower of the two pommels on a sidesaddle with a small amount of adjustability.

Leg up A method of mounting, in which an assistant stands behind the rider and supports the lower part of his left leg as he springs up off the ground.

Livery (stables) Riding establishment where an owner may keep his horse for a fee.

Long Reins Long, webbing reins attached to the bit of a horse's bridle and used in the animal's training.

Longe, to The art of training a horse by directing it around in a circle while on a long "longe" rein. This rein is attached to a cavesson (qv). Schooled horses may be longed as a form of exercise or during the course of teaching a novice rider.

Longe Rein A long, webbed rein used in the above action.

Manège A marked-out area or school used for the teaching, training, and schooling, of horse and rider.

Mare A female horse over four years old.

Martingale Ancillary item of tack, the purpose of which is to give a rider a greater degree of control.

Muck Sweat Condition of a horse when, through hard work or overexcitement, it has sweated to such an extent that its neck is covered with lather.

Mucking Out Daily stable chore involving the removal of dirty, soiled bedding and sweeping of the stable floor before replacing the bed.

Near side The left-hand side of a horse.

Neck-reining The art of turning a horse by using the indirect (qv) or opposite rein to the direction of the turn.

Neckstrap A simple leather strap buckled around the horse's neck used to give added security to a novice rider. Also refers to the strap of a martingale that buckles around the horse's neck.

Numnah A pad worn under the saddle, usually cut in the shape of the saddle. It may be made of felt, rubber or sheepskin.

Off side The right-hand side of a horse.

Overbend A horse that has arched its neck acutely, thereby bringing its head too far into its chest. Usually caused by a rider exerting too much pressure on the reins while urging the horse forward.

Paddock Fenced-in area of grassland in which horses are turned out. Generally used to denote a fairly small area.

Palomino Color of horse. The coat may be various shades of gold and the mane and tail white.

Pelham Various types of curb bit with a single mouthpiece to which two reins may be attached. Aims to combine the two bits of a double bridle in a single mouthpiece.

Piebald Refers to a coat irregularly marked with large patches of black and white. Pinto is an American term for piebald and skewbald horses (qv).

Pirouette A dressage movement in which the horse describes a circle in which the forelegs describe a small circle while the hind legs remain in the same spot, one of them acting as a pivot.

Points (of a horse) Names given to the different parts of a horse. Also used to describe the mane, tail, and lower legs.

Pommel The center front of an astride saddle. In some designs, the pommel is more pronounced.

Pony A small horse that stands 14.2hh or less.

Port A raised section in the center mouthpiece of some curb bits. It may be raised to a greater or lesser degree, thus affecting the severity of the bit.

Pull, to (mane and tail) The process of thinning the mane and tail.

Quarter, to Superficial grooming of stabled horses before taking them out for exercise.

Rein Back To instruct the horse to move backward. In order to execute the movement correctly, the horse must move back with the diagonal forelegs and hind legs moving in unison.

Reining Patterns Type of dressage test in Western riding in which advanced movements are executed.

Renvers A school movement also known as quarters-out, in which the horse moves along the side of the school, his hind legs on the track and his forelegs on an inside track.

Shoulder-In A two-track movement (qv) in which the horse is evenly bent along the length of its spine away from the direction of its movement.

Shy, to Wherein a horse jumps to one side having been frightened by a real or imaginary phenomenon.

Side Reins Reins used while training, to help position the horse's head. They are attached at one end to the bit and at the other to the girth or roller buckled around the horse's saddle or belly.

Skepping Out Stable management term used to describe the removal of droppings from the stable bed by putting them into a skep or skip.

Skewbald Refers to the coat of a horse irregularly marked with large patches of brown and white.

Snaffle (bit) Any one of a number of designs of bit that act on the corners or bars of the mouth. The bit takes only one pair of reins.

Spurs Small, metal devices (usually blunt) worn on the rider's boot to help reinforce the leg aids.

Stable Management The art of looking after one or more stable horses, including all aspects of their welfare.

Stall Old-fashioned stabling of horses, with several stalls usually incorporated in one building. Mainly found nowadays in large establishments, such as studs.

Steeplechase A horse race in which the horses gallop around a marked-out course which contains several, solid, brush fences. A steeplechase course refers to any course which includes such fences.

Stock A specially designed cravat, worn as part of a formal riding outfit, usually hunting dress.

Strapping The thorough grooming of a stabled horse.

Surcingle A webbing strap which passes around a horse's back and belly and is used to keep a rug in place. Show jumpers and jockeys often buckle a surcingle around their saddle as an added precaution against the girth breaking.

Tack Comprehensive term for saddlery (qv). A tack room is where tack is stored.

Tail Guard A piece of equipment made of leather, jute or wool, designed to completely cover the dock (qv) and protect this area of the tail. Frequently used when traveling.

Thoroughbred One of the most well-known breeds, the Thoroughbred, known also as the English racehorse, was bred for speed in the 17th century.

Trailer The transportation vehicle of one or two horses which is drawn behind another vehicle.

Transition The act of changing pace. A walk to a trot and a trot to a canter are known as *upward transitions*. A canter to a trot and a trot to a walk are *downward transitions*.

Travers Similar to a renvers (qv) except the forelegs stay on the outside track of the school while the hind legs move on an inner track.

Turn On The Forehand A school movement performed from a halt in which the hindquarters describe a circle around the forehand, with one foreleg acting as a pivot.

Turn Out, to To put a horse out to grass or turn it loose in a paddock.

Two Track School movements in which the hind legs follow a separate track from that made by the forelegs.

Vice Any one of a number of bad habits which may be learned by a horse. Unless curtailed when young, they are very hard to cure.

Volte A circle of 10 ft (3m) executed at a given point in the riding school.

Water Jump An obstacle, usually comprising a low hedge, behind which is a wide expanse of water. Used in show jumping courses.

Wisp A coiled or woven "rope" of straw used in strapping (qv) a horse, to massage the skin and muscles and to improve circulation.

Withers Point at the bottom of the neck of a horse from which a horse's height is measured.

Wither Pad A small pad, made of either felt or sheepskin placed under the front of the saddle to give added protection. In a well-fitting saddle, this should not be necessary.

Yawing When a horse continually opens its mouth and stretches its head outward and down to the ground in an attempt to evade the bit.

Yearling A colt or filly between one and two years old.

INDEX

ACKNOWLEDGEMENTS

Quintet Publishing would like to offer special thanks to
Bob Langrish, whose photographs appear on the
following pages:

2, 5, 6, 8t, 8b, 9, 10, 11, 12, 22, 23bl, 23bc, 23br, 25, 26b, 27, 28,
30, 32, 35, 36, 41t, 41b, 42, 52, 54, 62, 68, 91b, 92, 95, 98t,
100t, 100b, 101, 106, 107, 120, 123b, 125b, 128, 129, 130, 131,
132, 133, 134, 136, 141, 144, 145, 150, 154b, 156b, 157, 159, 161,
162, 166, 167, 170, 175b, 176, 178, 179, 180, 188, 192b, 195,
199, 201, 202, 203, 205, 210, 211, 212, 217